Polar Mariner

BEYOND THE LIMITS IN ANTARCTICA

Captain Tom Woodfield, O.B.E. Polar Medal

A reminiscence of voyaging for 20
years to the Falkland Islands, South
Georgia and the Antarctic

Whittles Publishing

To all those who sailed with me

Published by
Whittles Publishing Ltd.,
Dunbeath,
Caithness, KW6 6EG,
Scotland, UK
www.whittlespublishing.com

© 2016 Captain Tom Woodfield

ISBN 978-184995-166-1

Printed by Charlesworth Press, Wakefield

Contents

BUCKINGHAM PALACE

The Antarctic and its adjacent waters remain as hostile now as they were in the heroic age of exploration of Scott, Shackleton and Mawson, and throughout this text we are reminded of the even earlier courageous mariner explorers. Notwithstanding the challenges, the author's enjoyment of dealing with storms and ice, his love of the magnificent polar scenery and the abundant wildlife are clear.

It is interesting to be reminded of the early years of the British Antarctic Survey as the Falkland Islands Dependencies Survey, the marine support without which nothing could have been achieved. Captain Woodfield's account of that marine support and the exploration and survey work undertaken by the ships makes for fascinating and sometimes alarming reading.

This is an account of polar exploration, seamanship and human endeavour that is rarely found in this modern age and I am sure you will enjoy reading it.

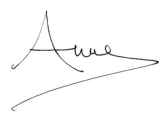

Acknowledgements

The initial guidance as to who to approach for this book to be published and advice on presentation I have to thank my Trinity House colleague and author Captain Richard Woodman. I am grateful to Adam Kerr for the loan of his journal of our first two voyages aboard the RRS *Shackleton* to back up my diaries. For their prolonged and untiring efforts to research photographs and copyrights I am most grateful to Ieuan Hopkins and Joanna Rae at the BAS Archives Department. My thanks also to Andrew Meijers head of BAS oceanography both for producing the oceanographic diagram of the Weddell Sea but also for bringing me up to date on such matters, and to Adrian Fox and Olivia Martin at BAS mapping department for producing the maps. Many thanks for the trouble some Fids have taken to send me images, particularly John Killingbeck and Tony Keville, although regrettably not all could be used. Thanks to my publishers editor Caroline Petherick whose skills tidied the manuscript and for her pointing out some anomalies for correction. It is hard to express sufficiently my gratitude to my wife, Ella, for her years of patience and support in my endeavours, for her interest in my maritime and polar world, and help sifting through the years of memories to decide on the contents of the book, and finally for laboriously deciphering and typing up my handwritten manuscript.

1 A Departure

I awoke to the occasional movement as the vessel came afloat alongside the old Public Jetty at Port Stanley. It was a bright summer morning in December 1965, but with a northerly of 60 knots blowing straight across the harbour, pinning us to the frail woodwork. I had been in command for a little over a year. A departure always excited me, and this time perhaps even more so because of the immediate difficulties. There were mutterings in the wardroom: 'It's madness to attempt to get off in this wind' and 'If we do, it will be hell outside', whilst on the bridge there were the usual calm preparations for leaving. With the majority of crew and scientists still in their bunks, we let go all lines, began hauling on the anchor cable that had been laid out on arrival to assist in just such adverse conditions, and used the engines and rudder to help the manoeuvre. The anchor slipped a little, but the chief officer on the forecastle knew from a gesture of my hand whether or not to heave. Gradually between us we worked *John Biscoe*'s head into the wind at right angles to the jetty. I applied some healthy power, and as we eased off the berth with the anchor weighed, the stern cleared the wooden structure by inches and we were away.

Three blasts on the whistle to wake the townsfolk and bid them farewell, easterly down the harbour then swinging north through the cable-wide[1] Narrows, the entrance to Stanley Harbour, with 70 knots of wind from right ahead. The red tin roof of the cathedral and the similarly clad multi-coloured roofs of the houses, some with their chimneys streaming smoke as early risers stoked their peat fires, Government House, with its driveway ablaze with golden yellow gorse but no ensign flying yet, and largest the bluff town hall, all disappearing from view as we passed from the inner harbour to the outer, Port William. Then a couple of miles easting, with severe gusts coming out of Sparrows Cove to port, where the hulk of Brunel's SS *Great Britain* lay ashore, and the sea, even with such a short fetch, dashing onto the Tussac Islands to starboard. Then a few moments of incredible movement as we cleared the protection given by the land and felt the full force of a Southern Ocean storm on our beam. The sea and swell were enormous, and from the bridge we watched the deck cargo and the scow[2] for any signs of shifting as they strained at their lashings, likewise keeping an eye on the boats moving uncomfortably in the davits. The stewards and cooks had been warned and had nothing remotely loose, but a couple of unfortunate sleeping scientists were thrown out of their bunks. Then, as we rounded Cape Pembroke with its lighthouse rising from the tussock-covered

---

1 A cable's length is one-tenth of a nautical mile, about 600 feet.

2 An open flat-bottomed boat, similar to a barge, used for transporting cargo.

headland, managed by Trinity House 7,000 miles away in London, and the sea pounding Seal Rocks close to starboard, almost obscuring them from view, I gave the order 'Steer 180, south'. We were heading *Down South*, those emotive words for all who venture to the Antarctic.

Breathtakingly beautiful and simultaneously daunting, it is the coldest and windiest continent on the planet and has the greatest average height of any. A continent that doubles its effective size in winter due to the formation of sea ice around its shores. A continent of extremes and anomalies, possessing a vast ice dome of several thousand feet in depth yet one of the driest places on earth, with only a minuscule amount of snow falling upon it, and that ice dome sandwiching a skim of water between itself and any rock that it sits on. A paradise of wildlife on its shores, where birds, penguins and seals frolic in abundance, yet the deep interior a wilderness whose sole inhabitant is the wingless fly. With weather turning in an instant from sublime to extremely severe, and where life or limb can be lost in a moment's indecision. A continent upon which the fulfilment of achievement and desperate frustration share the same stage as calm and gale, fields of sea ice and open water, mountainous glaciated ranges and flat snowfields to the horizon.

The plan had worked. As we rounded onto our course, the wind and sea came astern. Gusting above 80 knots, it stripped the top off sea and swell to create a sheet of spindrift streaming past us at main deck level. Then only a rather pleasant corkscrew motion and a feeling of being somewhat out of control as the sea from astern made steering difficult, but we were shipping little water and were in all other respects comfortable. In this exhilarating manner, with the barometer bottoming out, hopefully to rise, meaning a backing of the wind so that we would at least start with some shelter from the islands whilst the storm blew itself out, we made a flying start towards those southern polar shores.

My vocation

I was employed by the Falkland Islands Dependencies Survey (FIDS, also known as the Survey); its voyages were expeditions to relieve its bases in the Southern Ocean and Antarctica by bringing them new personnel and supplies. The bases were stations for scientific research, but in my first years their establishment also had an underlying political purpose. Our primary function was to support their very existence, then their scientific research, survey and exploration, and we carried out similar exercises ourselves offshore whilst we enhanced Britain's claim to sovereignty by our presence. The voyages were always a contrasting mixture of great endeavour and blissful scenic passage-making; ever present was the enthusiasm and camaraderie of our crew and the expeditioners. The combination of sea ice and ferocious weather which protects the shores of Antarctica and makes it at times such a fearsome place to approach provided continual challenges which I came to relish. There were many times of anxiety; being driven onto the rocks or becoming beset in ice and crushed for long periods, but also many of pure joy, gazing at the beauty of the land or seascapes, marvelling at the wildlife so tangible without fear of humans, or just feeling the satisfaction of achievement against the odds, whether on a day-to-day basis or on the completion of an entire voyage. I delighted in the company of my fellow travellers, be they humble mess boys, ship's officers, young scientists or eminent professors, when we shared the wonders and excitement of the polar world. I completed 19 seasons voyaging to Antarctica on three ships sailing from Southampton via South America and the Falkland Islands, having soon become passionate about my trade, polar seafaring, and fascinated by the magnificence, weather and wildlife of that southern continent. I became enthralled by the history of those who had first ventured to this desolate and demanding part of the planet, and almost incredulous of their achievements in appalling conditions, both under sail and ashore, with such poor equipment, clothing and food.

Earlier years

I ran away to sea aged 14 after an argument with my mother. I only got as far as the railway station, half a mile away. Confidence ebbing, I conveniently let two trains pass without boarding them before my father found me. He worked ashore in the shipping industry and was interested in all matters concerning the sea, and he had an enduring love of mountains and exploration; all of these he passed to me. He suggested that if I really wanted to be a mariner, which indeed I always had, I should go to sea school first.

After a tough but invaluable year at Warsash School of Navigation under the eye of the renowned principal, Captain Wally Wakeford, I joined the Port Line aged 16 for three and a half years as an indentured apprentice. I made several splendid voyages to Australasia; general cargo outward, and wool and refrigerated goods home. The ships, which carried 12 passengers, were excellent; they were well run, and the apprentices' maritime studies were well supervised. We spent up to six weeks on the coasts discharging and loading, and life could hardly have been better for us young men. Only one incident disrupted those heady years, with their mixture of hard work learning my trade, good fellowship, and the enjoyment of the sunny outdoors of Australasian sporting and beach life, mainly in the company of girls. Towards the end of my apprenticeship, outward bound to New Zealand I was put ashore delirious in Panama. In Jamaica I had swum in a hotel pool and fallen ill; the initial diagnosis was meningitis, but in fact it was only a tropical bug, and four months later I was back in England sitting my second mate's examination. Passing this in the summer of 1953 entitled me to keep a bridge watch on my own and thus begin the progression towards acquiring additional sea time and passing the examinations for first mate and then master.

An opportunity grasped

One Sunday shortly after I had gained my initial certificate, the good fortune from which my life has benefited began to kick in. During the afternoon my father read out an advertisement from the paper, adding, 'This will interest you, son.' It read, 'Full crew required for an Antarctic Expedition Ship – no polar experience necessary. Apply Crown Agents'. The opportunity to combine those inherited interests of the sea, mountains, and polar exploration was a chance not to be missed. I telephoned my employer on the Monday morning and explained that I wished to attend an interview for a junior officer's position aboard the Royal Research Ship (RRS) *Shackleton* rather than join my next Port Line vessel that day. To my surprise – for ship owners were tough employers in 1955 – they thought it a good idea and agreed that if I were accepted I could take leave from their employ for two years. I sat on a bench at the Crown Agents acting on behalf of the ship owners, awaiting my interview alongside an Adam Kerr. It turned out that we were both successful, he to become second officer, I to be third, and thus began a friendship that would last a lifetime. One of the interview panel was Captain Bill Johnston, who was to transfer from the RRS *John Biscoe*, the sole vessel then operated by FIDS, to command its newly acquired second vessel, to be named the *Shackleton*.

In 1943 Argentina had made a definitive and substantial claim to the segment of Antarctica and the sub-Antarctic islands lying to the south of Tierra del Fuego, bounded approximately by the extreme east and west longitudes of the Argentine and Chile – the same area that the UK laid claim to, and called the Falkland Islands Dependencies. In response, the British dispatched a secret naval force, Operation Tabarin, to prevent an Argentinean/German collaboration taking control of the Drake Passage, the sub-Antarctic island groups of the South Shetlands and South Orkneys, and the waters that lay between them and mainland Antarctica. After the war's end in 1945, that operation matured into the FIDS, a scientific survey with a strong political emphasis. Further stations were gradually established and eventually a dedicated vessel, the first *John Biscoe* (a new vessel of the same name was to be built in 1956) was acquired late in 1947.

Several years later, in 1955, we were to join the vessel to be renamed the *Shackleton* in Frederikshavn, Denmark, where she was being refitted for her new role. It had become apparent to the Survey that a single vessel for year-on-year operations could not meet their requirements and that operating in those southern waters on her own would be unsafe. That vessel, the first *John Biscoe*, a former wooden boom defence vessel of about 1,000 tons displacement, could scarcely carry the stores and personnel for the resupply of the Antarctic bases, let alone the materials required to build new stations, without several return trips between them and the Falklands. These time-consuming voyages across the often turbulent Drake Passage would remove her from the field of operations in mid-season, when she was required for her secondary role of supporting detached survey parties and generally assisting movement of personnel, stores and dogs between bases. Heightened roles were also envisaged of delivering and assisting with the build of several new bases and supporting scientific work.

Safety of the personnel carried and that of the life-supporting base stores had also been a prime consideration in the decision to acquire a second vessel. Safety would be greatly enhanced by one vessel being able to go to the aid of a sister ship should she become stranded or holed in the uncharted and ice-filled waters south. Furthermore the loss of a single ship would wreck an entire season's programme or worse, whereas if there were two ships sharing relief and other duties, the loss of one would have less impact on the overall operation. The introduction of a second ship would also greatly improve the possibility of seaborne science being undertaken in line with the growing emphasis on marine science, such as biology, oceanography and the geophysical investigations relating to the movements of tectonic plates and the theories relating to Gondwanaland. Politically, it was obviously preferable to have as many activities as possible in the disputed areas to exercise and display the British claim to sovereignty – and what better than another ship plying those waters, literally flying the flag?

The choice of a second ship

Building a vessel from scratch – evaluating and drawing up the requirement, trawling for naval architects, preparing the design, going out to shipyards for tender, tank testing and finally the build itself – is a lengthy process. Too lengthy when government policy was to improve our political footprint by way of additional stations and to increase our scientific output with safer and more capable ships as fast as possible. Plans for a replacement *John Biscoe* were already in place, she to be purpose-built.

An icebreaker is by definition a vessel full of engines giving it the propulsive power to break ice, and having little or no space for cargo. Large ice-strengthened cargo carriers are strong enough to withstand ice when following an icebreaker, but have little capability to do so on their own because of their low power to weight ratio. On well-worked established routes through ice in the Baltic and the Arctic, a government icebreaker leading a convoy of commercial cargo ships paying for the service was a well-proven concept; but in heavy polar ice, particularly in at best poorly charted waters of locations visited by perhaps by just a single vessel, icebreaker support was neither a practical or financially sound proposition. Nor was it financially realistic to build both an icebreaker and an ice-strengthened cargo vessel to achieve the same objectives, although the Admiralty entered the discussion for a while with the idea of the navy operating an icebreaker in support of the Survey; but this came to nothing. Thus a composite vessel, powerful enough to work polar ice on her own, having ample cargo space, abundant cabin space for expeditioners, small enough to work pack ice – length being an important criterion for this – and of shallow enough draught to enter the tight anchorages of the Antarctic Peninsula where most of the bases were located, was the FIDS' requirement. The Russians, Finns, Americans and Canadians had both breakers and large ice class support vessels, but almost nothing to suit the FIDS requirement. The Danish Lauritzen Lines had their Dan ships on the Greenland trade; these were eminently suitable, but none were for sale and at that time it was not thought an economic proposition to charter them for the nine-month seasons south.

Ever since polar exploration had begun, British expeditions had simply acquired the most suitable vessels available at the time, these invariably being sealers or whalers, with those functions their prime design criteria. They were strong and seaworthy enough to reach their far-flung destinations but not actually designed to work ice. Nevertheless, through their inherent strength and their lack of power so that they could never be forced too hard in pack ice, together with careful handling to avoid damage from impact amidst loose ice at speed, they served many expeditions incredibly well. The navy had also reinforced a number of its ships, particularly for the 19th-century expeditions to the North-West Passage. Two 'bomb' ships,[3] *Erebus* and *Terror*, on separate expeditions commanded by Sir James Clark Ross and

---

3 Built between 1680 and 1856, mostly to bombard shore fortifications with mortars, their construction being very strong to accept the recoil when the mortars were fired.

Sir John Franklin, made miraculous voyages both north and south, their sturdy build suiting them well to polar exploration, until their mysterious loss together under Franklin during his 1845 Arctic Expedition to find the North West Passage.

Eventually, brokers found the pretty little *Arendal*, a year-old Swedish Baltic trader of 980 tons displacement, classed for working the ice of that region. One hundred and eighty feet in length, she had mainly three eighths of inch plating, with frames set 24 inches apart. Two holds forward with sliding metal hatch covers into which were incorporated two cranes, and tween decks[4] within. A small stores hatch on the forecastle, and the bridge, and accommodation for a small crew aft. Four watertight bulkheads to main deck level, some double bottoms[5] but not beneath the engine room, which was also aft. She was powered by a single MAN diesel of 800 horsepower driving a Kamewa variable pitch propeller made of austenitic steel, a very hard material which would accept the impact of ice, and whose four blades were individually detachable so that they could be replaced if damaged.

---

4 Decks situated between the lowest in the ship and the main deck.

5 Shallow tanks formed in the bottom of a vessel utilising the hull plating and an interior deck of plating called the tank-top. Either containing ballast or fuel, or left empty and providing additional means of flotation were the hull to be ruptured.

2 An Unstable Beginning

Nearly gone before we began

During December 1955 I travelled to Frederikshavn, where the *Shackleton* lay, with Adam Kerr and a small number of crew. Arrangements had been made by the Crown Agents in accord with some code of their own, which dictated that Adam and I travelled first class on a North Sea ferry to Esbjerg yet third class on the train across Denmark to Frederikshavn. Our sea passage was enhanced by a strong gale that sent all passengers to their bunks save a gentleman from the BBC, a dwarf joining a circus in Copenhagen, Adam and myself. The ferry was on her last voyage before laying up for the winter, and the barman insisted we help clear his stock; we had a hilarious time and saw little of our first-class cabins, which was to be regretted the following day when trying to get comfortable to shake off our hangovers on the wooden slatted seats of our third-class railway carriage.

Frederikshavn was a delightful place to visit, but proved to be colder at times than some of the Antarctic stations that we were to visit later. With the exuberance of 22-year-olds, we enjoyed our few weeks there, and my memories are of beetroot soup almost every night to start dinner, warm candlelit coffee shops with delicious pastries after the day's work, and having to go to the public baths for a shower and then return in 20 degrees of frost, because flower pots were kept in the only bath in our guest house. I also remember the landlady's daughter – and I remember removing the carpet rods from the stairs to her room above ours prior to one of our number creeping back down to his own in the early hours, which resulted in a thunderous and, for him, embarrassing commotion.

We had arrived in Frederikshavn to stand by the ship as the modifications to make her suitable for service in Antarctica were being completed. The hatch covers of the after hold had been removed, and the bridge and adjoining accommodation moved forward over that space. That hold was partly fitted with further accommodation, storerooms and a freestanding fresh water tank. What remained of it was incorporated into a single hold for cargo. More cabins were built aft of the bridge, and extra lifeboats were housed on her elongated boat deck to provide for the new complement of 30 crew and 32 expeditioners, increased from the original total complement of 14. Water ballast tanks were converted to fuel tanks to meet the required steaming range of 10,000 miles, and a fifth, additional, watertight bulkhead was fitted amidships to main deck level with a watertight door in the 'tween deck.

A few days before we sailed for Southampton, the shipyard personnel carried out a heeling experiment, or inclining test, to ascertain the vessel's stability. The weather was poor, very gusty, and the decks slippery with the dusting of frozen snow. The vessel was hauled into mid-harbour but because of the wind could not be freed from the restraint of her mooring lines, as is the norm for such a trial. Whilst a plumb line was hung into the hold above a measuring baton, known weights were moved across the deck to heel the vessel, and the distance the plumb moved with each shift of the weights was measured. Thus the stability criteria were obtained from which the stability book containing the stability curves and figures was to be produced. As with all ships, these are used continuously throughout a ship's life when loading, taking fuel, water or stores, to position them correctly or dictate any ballasting or trimming. The aim was to keep a ship stable and upright without being too stiff in a seaway and therefore too violent in her motion, yet not too tender and thus lazy of roll so that she lolls too long over to one side before returning to the upright.

After all the modifications had been completed, the voyage to Southampton started with us heading straight out into a gale of some severity, in which her lack of power to maintain a reasonable speed into heavy weather became immediately apparent. We took two days to progress through the Kattegat and round the northern point of Denmark. There was no question of heaving to in such conditions; she could barely keep any headway. On board for the passage were the aged senior partners of our appointed naval architects who had overseen the refit and the modifications. One had a walking stick with which he prodded a piece of structure as he passed, exclaiming what a fine vessel she was. But we began to think otherwise; our new ship, in which we were to sail on the adventure of a lifetime, rolled so alarmingly when wind and sea came onto the beam that we feared she would capsize. Eventually the weather eased, and we made our way down the North Sea with most minds turning to the loading in Southampton and our planned departure from there just before the Christmas holiday.

Not so the chief officer. Tom Flack, in his late thirties, had gained quite wide experience, having taken a variety of positions around the world. He appeared to be very much an academic, holding part of an extra master's certificate, whilst his ability as a practical seaman was not possible for us to ascertain at this early stage of the voyage. He pored over what information he had, and made calculations of his own. He then consulted the master, Bill Johnston, who had actually not shown a great deal of concern during the passage. This was to be our first indication of Johnston's imperturbable character. Flack then went ashore in Southampton, with Johnston's permission, to air concerns regarding our stability to the chief surveyor, Captain Stephenson, a tall, approachable father-like figure at the Board of Trade shipping office. Captain Ron Freaker, a diminutive man with a terrifying manner for young officers taking their oral examinations, was also a senior surveyor there, as was a Mr Jim Cox. Prior to joining that department, Freaker had been master of the RRS *William Scoresby* during the late thirties' investigations in the Southern Ocean; he clearly relished the task of ensuring that everything in this foreign-built vessel was up to the safety standards of the Board of Trade before we sailed south. He was diligent and painstaking, indeed fastidious, and it seemed to us that he was determined to wreck our timetable for loading and departure with all the modifications to equipment and additions to stores that he imposed. We only just escaped having to replace our open aluminium lifeboats.

On the question of stability, he ordered a further inclining test. We moved from our loading berth, Quay No. 37 at the outward end of the docks, to a quiet inner basin. There, under stringent but much more appropriate conditions, we went through those procedures again, this time with many observers, including the Lloyd's Register of Shipping surveyor, because we were to be classified and insured by Lloyd's.

The results were alarming. Had we been hit by a large wave on the beam when rolling it was more than likely that we would have capsized. And at Southampton we were to ship both our scow and our workboat, each weighing nearly four tons, on the upper deck, which would have made matters worse. Further we had been overbooked with cargo, much of which was building materials for new base stations, forcing us to carry much of it as deck cargo. The situation was dire – and, with the usage of fuel from the double-bottom tanks on passage reducing our stability to negative, potentially catastrophic. Freaker and Cox placed us under a detention order, to be lifted only when corrective measures had been taken. Just over 100 tons of solid ballast would have to be placed as low as possible amidships, to be followed by a further inclining test. The loading of the cargo and our own stores, the embarkation of the expeditioners, and our departure for Montevideo, the Falklands and Antarctica, were all delayed until after Christmas. Other than on Christmas Day itself, pig iron was loaded into the hatch and transferred by hand to a location beneath the water tank that had been added in Denmark; an ideal housing for the ballast, but difficult to access.

Of the 12 ship's officers, some could get home for Christmas and Boxing Day whilst the rest had to remain on duty. We decided therefore that we would all remain on board together for the holiday, a good indication of the camaraderie that was to build during the voyage. We enjoyed our Christmas lunch together, and someone had, in true merchant service fashion, telephoned the Southampton Nurses' Home and arranged a party with some of them on board in the evening. A matron escorted 12 delightful young ladies to join us, which resulted in a great party, the captain entertaining the matron. Several romances ensued, and two marriages followed a year or so later.

Southampton was to be my home port for the next 20 years, and even on this first visit I began to develop some lasting friendships within the maritime community. Both our agents, McGregor, Gow & Holland, and Thornycrofts the ship repairers served us well and remained engaged by us throughout my time. The harbour authority, shipping office, customs, nautical surveyors, ship chandlers and police tended to our needs, and their personnel became friends then and on all subsequent summer visits to the port. The ship, registered in Stanley, F.I., had the title Royal Research Ship, which appeared on our register – as did her owner, Her Majesty the Queen – whilst we flew the defaced blue ensign of the Falkland Islands. We were few, young and enthusiastic, and regarded our voyages as exciting. All these things contributed to our being treated exceedingly well, and great relationships were formed. The owner also paid her bills on time! Our annual farewell party before each voyage, bringing together our small London staff headed by our director, and our Southampton friends, became a notable port event, but not on this hurried winter visit. Though we did manage to fit in a naming ceremony, performed by Lady Arthur, the wife of the incumbent Governor of the Falkland Islands, in front of her husband and some eminent Falkland and Antarctic guests, including Sir Miles Clifford, a former governor, Sir James Wordie, and the son of Sir Ernest Shackleton, Edward, an MP (later Lord Shackleton), whom I met for the first time, and who was to influence and be involved in my life from then on.

Festivities over and ballasting completed, lifesaving and safety measures were put in order in accordance with Captain Ron Freaker's requirements. He, we soon came to realise, had a heart of gold. Everything he required to be rectified was entirely for our benefit. He maintained an interest in our ships and activities well beyond his retirement.

3 RRS *Shackleton*: First Voyage

Lasting impressions

The loading of the cargo and stores was completed quickly once the inclining test had proved our stability satisfactory, although we still had a condition placed upon us to ballast any double-bottom fuel tanks as they became empty during the voyage. The expeditioners were embarked, and at last we sailed, but again it was into poor weather. Down the Channel towards Ushant a full southwester came in and Johnston decided a quiet anchorage made more sense than little and uncomfortable progress in mid-Channel. We anchored less than a cable under the lee of Berry Head, weaving our way through many other ships taking shelter.

To those of us from large ships, the philosophy of nursing a small ship and crew, working with the weather rather than fighting it, was interesting and my first simple lesson in seamanship from observing Bill Johnston. He was however to prove a hard taskmaster. He was a tall, gaunt Ulsterman, brought up in the coastal trade. He had experienced a tough war in rescue tugs based on Gibraltar, assisting Malta convoys, and had then joined the Falkland Islands Company's ships as master. Sir Miles Clifford, who had admired his expertise, recruited him to FIDS. The governor represented the Colonial Office, the body responsible for us. Johnston was a cold, rather aloof figure, who chain-smoked Players Perfectos Finos, and drank much pink gin. Not once, though, in my nine years associated with him, did I see him the worse for wear, not even in the Falklands where the hospitality of the inhabitants made it difficult to stay sober. He occasionally showed a dry sense of humour, though usually at the expense of others. He did not suffer fools or incompetence, and was severe if you were not carrying out his instructions to the letter. He always wore uniform, usually with cap, even in the severest weather. He frowned upon any dressing down and even showed a dislike, but tolerated, my wearing a black silk scarf instead of a tie, when passage-making at night. A favourite quote of his was 'You don't have to be scruffy to be tough'. His authority was absolute, but his style of command, which extended beyond the ship to the bases and areas in which we were working, was extreme. He never discussed his plans or tactics. He never aired his concerns nor shared confidences. He was phlegmatic, with apparently no nerves. We used to joke that if you stuck a pin in him he might later consider saying 'ouch'. His well-kept secret, however, was that he was seasick, although some sharp-eyed lads noticed that he was not around much during the first day of a passage if it was rough unless he thought it absolutely necessary. Above all, though, and most importantly, was that he was a fine, safe, and adventurous seaman, with an uncanny eye for a safe passage in unchartered waters, from whom I was to learn a great deal.

An uneventful passage was made to Montevideo with a short visit to the Cape Verde Islands to top up with water and fuel, mainly for the purpose of maintaining good stability. During the passage we familiarised ourselves with the ship and settled into routines. We had none of the teething problems of a new vessel, for she had already operated for a year in the Baltic. However on the magnetic equator, at about latitude 5 degrees south, we did swing ship to calibrate the magnetic compass. Taking compass bearings of the sun on different headings as the vessel swings through 360 degrees and comparing them with the true bearings extracted from the Nautical Almanac establishes the deviation of the compass on various headings. This is then refined by adjusting the Flinders Bar and Kelvin Balls.[6] The change to the *Shackleton's* original errors would have resulted from a change to the magnetism of the ship consequent on the large amount of structural work that had taken place in the shipyard. This was a first for me, because a professional compass adjuster usually carries out the task prior to sailing, but this was far from being my last 'swing'; we were to discover that when working heavy ice the effect on the ship was the same as that of being hammered in a shipyard, changing her magnetism and consequently its effect on the compass error. We had therefore to devise methods of evaluating the deviation whilst in high latitudes, where the other compass error, variation, from the magnetic field of the earth, was high, not nil as at the magnetic equator. We created many transits at places we frequently visited, and whose true bearings we had established on sunny days, in order to enable us to check the compass when required. We had a gyrocompass, but during this initial season, and on most subsequent ones, it repeatedly failed, once for an entire nine-month voyage. I therefore developed an obsession with the errors and corrections of the magnetic compass.

Montevideo, at 36 degrees latitude south, in the mouth of the River Plate, was an exciting and friendly place to visit. It was hot in midsummer, the air filled with the aromas of frangipane, bougainvillea, barbecued meat and red wine. Scruffy in a Latin unkempt way, with buildings in disrepair and broken pavements, it had lovely beaches and outdoor way of life. Our day and a half there were spent taking water and fuel. It was also our last chance to stock up on fresh fruit, vegetables and meat, none of these being available in the Falklands except mutton, which was abundant.

The voyage of 1,000 miles from there to the Falklands saw a distinct drop in air temperature. There are two closely parallel but opposing currents off the east coast of Argentina, the south-going warm, the north-going Falkland Current cold. The latter had carried an iceberg onto the English Bank in the mouth of the River Plate in 1936. I never met ice until some few hundred miles south of the Falklands, but a good lookout was always kept for it on this passage, particularly in fog. This formed frequently when the cold air above the Falkland Current blew over the adjacent warmer water and condensed.

Stanley and introduction to the Southern Ocean

We were well behind the schedule for a southern season when we arrived at Stanley late in January. FIDS had its rear base here; at the back of Government House were a handful of

---

6 Kelvin Balls are soft iron spheres either side of the magnetic compass which can be moved to obviate anomalies in the ship' magnetism. Likewise the Flinders Bar, immediately forward of the compass.

offices where a small team led by the Survey's Secretary (Secfids) co-ordinated activities south, liaising with the governor and London HQ through a wireless office nearby. This provided a communications centre for the ships, bases and shore staff. Near the jetty we also had a stores depot. The Fids (as the expeditioners were known) were kitted out with their 'southern' clothing and gear, whilst stores, cargo and mail were transferred between ship and shore so as to stow our load in accordance with the order of bases we were to visit on this single voyage south, in our shortened season. The primary role of the vessels was to resupply the several bases with their annual stores and change over the personnel, after which we were to support their operations. We therefore expected to build refuge huts as outposts for extended survey runs from the bases, and to land men, dogs and stores at remote locations to establish camps from which further fieldwork could be undertaken. On this voyage, however, in conjunction with our sister ship, we carried the building materials for the erection of two complete new stations, including generators, food, fuel and equipment sufficient for a year's occupation. Another first for me as we sailed out of Stanley Harbour was not just to have the decks piled high with drums of petrol and diesel oil, gas cylinders and sledges, but also to have fresh carcasses of mutton strapped to the rigging. After a couple of days at sea they were perfect for delivery, being salt cured and frozen.

The Drake Passage, lying to the southward of the Falklands and South America is a tempestuous place, the passage of depressions through it being almost relentless and the swell they create rarely subsiding, and averaging between four and five metres. Twenty years later this stretch of water was to throw at me the most tremendous family of ocean storms of my career, but on this occasion, sailing between depressions, it was relatively quiet. At 58 degrees south a faint speck of white appeared on the horizon – our first iceberg, always an amazing sight and wonderful experience, whether a giant flat-top tabular broken away from an ice shelf or a sculptured deep blue fragment from a glacier. As it drew nearer, one became aware not just of its immensity but also of the chaos of surging, breaking, and swirling sea that its static bulk created in the incessantly moving ocean, this maelstrom in turn trapping fish, squid and krill, providing a feeding ground for a multitude of birds.

At about this latitude we crossed the Antarctic Polar Front or Convergence. This is where the cold north-going surface waters surrounding the Antarctic meet and sink beneath the southward-flowing more temperate waters. Also in this area is an upwelling of water from the circumpolar current created by the almost constant westerly winds. The sea thermograph[7] fell abruptly by about 5 degrees, showing that at sea level there is very little mixing between the two adjacent water masses. The relevance of these oceanographic phenomena, the instant change to cold water and the upwelling, is to the mariner twofold. Firstly, the colder water allows ice to survive longer than in the warmer, thus increasing greatly the possibility of ice being met in the open ocean. Secondly, the bird life increases dramatically. The cold water, initially about 2 degrees centigrade, supports krill, which are fed by the nutrients and diatoms carried to the surface by the upwelling of the circumpolar current. Around the ship, immediately on entering the colder, richer water, giant wandering albatross that had occasionally been sighted

---

7 Device which records the sea temperature on the bridge, from an instrument in the hull.

singly since about the latitude of Rio de Janeiro at 23 degrees south, now arrived abundantly. Rarely was the ship then without these graceful creatures. They were joined at the other end of the bird spectrum by the tiny Wilson's storm petrels, both riding the air up-draughts from the plunging bow or feeding on whatever was churned up in the wake astern.

Admiralty Bay

As we approached the sub-Antarctic islands of the South Shetlands a wonderful variety of land-nesting seabirds joined the ship – terns, skuas, shearwaters, shags and gulls. Turning into our first southern harbour, Admiralty Bay on King George Island, we also entered our first pack ice, blown into the 12-mile-deep enclosure from the Bransfield Strait. Working multi-year pack ice,[8] some three feet out of the water with mostly considerably more beneath, seals upon it basking in the sun, and penguins hopping on and off the floes joining the array of wildlife, was fascinating. The ice was variously pristine white with snow cover, bare and deep blue, light blue to green with algae, and dirty brown with decay or the detriment of seals and penguins. The rugged island 1,800 feet high close to hand, almost encircling us, was heavily glaciated, crevasses galore in the tumbling ice fields ending at the water's edge in fractured ice cliffs, interspersed by craggy rock buttresses, mountain peaks and nunataks[9] rising inland from snowfields. Every ray of light, whether of direct sun or a shaft filtered through cloud, burst into a spectrum of colours as the ice crystals of snow, pack ice or glaciers acted as prisms. The low sun cast long dark shadows that contrasted with the pinks, blues and greens created by the algae within the floes or ice cliffs and the rose tints of the glacier and shelf faces when struck directly by the sun as it skimmed along the horizon at the beginning and end of our near 24-hour day, it being midsummer. What a glorious sight, a natural wonderland, and a realisation by us that Antarctica was not all black and white. This was my first sighting of Antarctica in all its glory, yet to be dwarfed later by even greater panoramic extravaganza.

On arrival we anchored within the pack ice, but it was too consolidated for us to lower the boats and work cargo, yet too hummocked to carry it across by sledge. There was also always the possibility that the wind would change and the ice would loosen or move out, stranding us with our loads, or worse. Though it was satisfactory for us to take a first short polar walk around the ship to get some photographs. Twenty-four hours later the pack did loosen and we put our boats down and got ashore. We landed the stores on a shingle beach adjacent to the base hut that was littered with a variety of whalebones. These, and the occasional iron try pot in which whale or seal blubber had been rendered down, bore testament to the whaling and ruthless slaughter of seals in the early 19th century; fur seals were taken for their coats and any species for their blubber.

Our method of cargo delivery at such anchorages as this base afforded was in the scow, a 32-foot open wooden barge, strapped alongside our sturdy motorboat. Fairly straightforward in good weather with no swell and no ice; but that was not often the case. The anchorage, one

---

8 Sea ice formed over several years.

9 Isolated rock peaks protruding through glacier or ice sheet.

of the best, had good holding in mud, but we were harassed by pack ice moving in and out on the wind from the Bransfield Strait. Its weight would drag us around, and threaten the boat and scow working alongside. Frequently the landings were clogged with floating brash ice[10] or bergy bits,[11] or the same left stranded on the beach by a falling tide. The weather was changeable, gusting winds producing breaking waves that, together with any swell, made work difficult amongst the ice. All these circumstances we began to master, and they became the norm. We usually took large items such as tractors and generators ashore by forming a platform across both motorboat and scow, upon which they were placed. They were then driven off or hauled down a ramp of wooden deals[12] that we carried. Fids were the main source of power for landing cargo: we had never heard of health and safety.

Before leaving Admiralty Bay we shifted anchorage and tied up stern to a glacier snout where the base Fids had created a dam amongst the rocks immediately at its foot. We ran out a hose and pumped good fresh water aboard. The glacial scouring provided a danger-free approach, and the debris good holding for the anchor. However the heaped moraine at the sides of the glacier continued out into the sea forming shoal water, and had to be avoided. With our complement we were always short of water, and tapping these glacial runoffs was a favoured and frequent method of filling up our tanks at a variety of sites.

The Banana Belt

We progressed south to the UK stations of Deception Island, Port Lockroy, Arthur Harbour and Argentine Islands, and searched for a new base site on the Antarctic Peninsula, calling at several unexplored locations on both the mainland peninsula and the offshore islands. The scenery became grander and more beautiful. The mountains of the mainland rose to 12,000 feet. The glaciers descending from them, forming at greater altitude and further inland, were very much longer and wider, with higher fissured and fractured cliff faces where they terminated and entered the sea, frequently calving. The coastline was mainly of ice cliffs interspersed by higher solid rock buttresses. Yet later, after we had voyaged further south, this region of the northern peninsula with its more moderate temperatures became known to us as the Banana Belt!

The sea ice we continually met varied from one-year-old to multi-year. The younger ice would have broken out in the spring from an adjacent bay or fjord, of which there were many on this coast, where it had established and remained over the previous winter. Usually it was no thicker than a foot, very even, and often in pieces up to a mile across. The older ice had formed in more open water, then become fragmented by swell. Driven before the wind, it would have consolidated and rafted as it came under pressure against further ice, grounded icebergs or land, and formed into an ice-field. As such, more severe rafting would take place, as in this state it became static in a further winter's freeze. Fragmenting during the following

---

10 Fragments of floating ice less than six feet across.

11 Floating ice less than 15 feet above water and about the same distance across. Minor icebergs.

12 Baulks of timber.

spring thaw, the ice could go through this cycle many times, producing multi-year ice with floes up to 40 feet thick with varying degrees of hardness within. A further feature of this ice is that it can contain vicious rock-like cores of glacial or shelf ice fragments. The ice of this latter type in our vicinity would undoubtedly have been formed in, and then borne out of, the Weddell Sea by the circulatory current, there to flow past the tip of the peninsula, then driven by the wind into the Bransfield Strait and beyond.

Entering this ice for the first time, the rules of engagement were drilled into us, as years later I drilled them likewise into my young officers: reduce speed; do not get embroiled in ice until there is no alternative; go round every piece, large or small, vast floe or ice-field, until you can no longer steer a sensible course to your destination. This was safe practice, and good training for those learning to handle the ship. Do not collide with it unless there is no alternative, then adjust speed and hit it with the stem. Avoid a glancing blow on the shoulder, where the arrow-form of the bow ends and the full body of the hull begins, it being the most vulnerable of areas. If in doubt about one's ability to avoid a piece, alter course to hit it head on, doing whatever one can with the engines to reduce the impact. Remembering that water driven over the rudder from the propeller is what steers the ship best, therefore having possibly thundered astern to reduce speed on impact, it often pays to go ahead again momentarily to get that flow of water across the rudder and gain finer control before colliding. We soon realised that the master kept himself in reserve for the most difficult conditions and that we young deck officers were expected to handle the ship in a manner few deep sea shipmasters ever had the chance to.

Deception Island

We then headed towards Deception Island in the South Shetland Islands, which contains the best known and most used harbour in the northern peninsula and off-lying islands. With British, Argentinean and Chilean stations on its shores, it is the ugly duckling amongst the scenic beauty of the other base sites, yet perhaps one of the most interesting places to be visited in the region. Volcanic, it is the second largest crater island in the world. It had been discovered by Captain William Smith in January 1820, in the 250-ton brig *Williams* during his fourth voyage to the South Shetlands. These voyages were of a commercial nature, round Cape Horn between Buenos Aires and Valparaiso. Yet he chose on each occasion to strike well south of the Horn in the ambitious hope of finding a southern continent. His courage, repeated determination and outstanding seamanship, amongst ice, and in virgin waters, cannot be lauded sufficiently.

On his first voyage there, in February 1819, he made a landfall on the north coast of Livingstone Island, naming it Williams Point. He then bore away northward. On another voyage during the southern winter of that year he saw no land, being held off by pack ice, but on his third voyage the following spring, he twice ran the length of the northern shores of the group, discovering all the major islands including Deception. As he departed north and west on this voyage he sighted the magnificent island that now bears his name. Only 15 by 4 miles in size and lying some 20 miles to the westward of the other islands of the group, it rises sheer out of the sea to a tricorne of peaks, the highest at 6,900 feet. It has precipitous bare rock faces which contrast dramatically with its steep snow slopes, and is difficult to land upon.

It was, however, the 22-year-old American sealer, Captain Nathaniel Palmer, who in November 1820 in the sloop *Hero* is thought to have first entered the harbour at Deception Island and seen its potential. The horseshoe-shaped narrow crater rim, averaging 1,500 feet in height, encloses a circular bay four miles in diameter with a central depth of 88 fathoms, and the adjoining very much smaller and shallower Whalers Bay, a secondary crater on which our base was located. The only entrance to these bays is through Neptune's Bellows, named by the early sealers on account of the gusting winds so often experienced there. A cable wide, it is as dramatic as its name, with 300-foot-high vertical cliffs to starboard as one enters, with strata of varying volcanic hues of sulphurous yellows, oranges and brick reds. Separated from these cliffs by only a few feet is the remarkable dark basalt Pete's Pillar, an Old Man of Hoy-type rock stack of 150 feet. The apparently usable water of the entrance is halved, for to port lies a hidden central danger, Raven Rock, with only six feet of water over it, and beyond it more foul ground.

Later in my career, my own rather unorthodox route of entry was to steer for Pete's Pillar until 400 feet from it, then alter onto a course which bore towards a distinguishable rock point within the harbour which kept me a constant 500 feet off the cliffs in the best water, and safely past Raven Rock. Initially steering directly for the towering shore before altering course, was slightly dramatic and unconventional, but I felt by doing that I had more control in gusting winds and in better water than on a long single course approach in unnerving shallow water, attempting ultimately to achieve, and maintain, that small distance off the cliffs correctly, especially as any leading mark was often obscured by mist or a lump of ice. It also gave me a better view, as I approached, of the inner entrance and harbour. The reason for wanting this was that by the time I had command, I could see, beyond Raven Rock and the foul ground on the more gently rising shoreline opposite the high cliffs, the wreck of a Salvesen whaler, the *Southern Hunter*. She, having entered the harbour in poor weather to see if any compatriots were sheltering there, found none, took a round turn in Whalers Bay and proceeded out to sea again. But vessels entering and leaving are obscured from each other on the dogleg course needed to enter and exit Whalers Bay until the last moment by the cliffs. She had met an inward Argentine supply ship in centre channel, altered course to starboard to avoid her and hit Raven Rock. Since that occurrence most ships blow their sirens when entering or leaving. I also took the precaution of making my unconventional and exciting approach.

On the edge of Whalers Bay the base hut stood amidst the remains of the derelict whaling station which it had been built from. The shoreline was fascinating in that there were many hot springs, and at low water much steam would rise from it. Inland upon the ash and snow slopes there were fissures in the outcrops of rock gushing more steam and smoking fumaroles. The snow slopes, to a large extent despoiled by ash, were occasionally relieved by red volcanic outcrops, but the general appearance of the island in mid-season was of a dull, depressing nature. The prolific and varied birdlife nesting among the rocks and cliff faces were a welcome counterbalance to the austere landscape. They appeared to relish a bathe in the warm springs, and the pintado petrel in particular loved to gorge themselves on the parboiled krill along the tideline. In 1921 there were reports from catchers in Whalers Bay of the seabed subsiding and paint being seared off their hulls during volcanic activity. Years later we were to have our own volcanic dramas, but on this first visit nothing worse occurred than the ship frequently dragging in high winds and Adam and me falling down a crevasse.

When the main discharge of cargo was completed the ship would invariably remain at a base for a while to provide support for the shore activities, and did so on this occasion. The ship's engineers would assist with the base generators, the wireless officer with the radios; the Fids aboard and crew would help generally in sorting and stowing stores – and all of us when required, which was often, with building works. The presence of the ship providing food, hot showers, the occasional evening film show, and the conviviality of some fresh faces with whom to share a beer, was an important part of our base visits. We three deck officers only stood anchor watches in the severest weather; the quartermasters and able seamen manned the bridge whilst at anchor, monitoring the ship's position by bearings, transits of shore features, and from distance rings on the radar set against the image of the shoreline. Watching out for the welfare of the boats and keeping in radio contact with them as they plied between ship and shore was a further duty for them. Tom Flack, the chief officer, preferred to 'keep' ship, which meant that Adam and I, when off duty, had time to follow our own pursuits. Adam's developing passion was surveying, whilst mine was to climb every piece of rising ground and see wildlife. We joined forces to support each other's efforts, mostly surveying from the launch or enjoying a hike. Our main problem off this base was dragging anchor, the result of a combination of poor holding on a steeply shelving ash seabed, and the sudden onset of frequent offshore gales, one lasting for three days. Yawing prior to dragging on that occasion, it became urgent for us to clear the scow, full of cargo, and motorboat, that were lashed together, from the ship's side. Our practice was to transport Fids from the ship, sitting atop the cargo in the scow. On this occasion, as we made for the shore the scow began to be swamped. The Fids scrambled across into the motorboat, already packed with personnel, as we cut the lashings of the sinking scow, which had begun to drag the motorboat gunwale beneath the water. The scow freed, and no longer weighed down by some 30 souls, lurched and shed some of its cargo, but remained waterlogged. We were able to tow it to the beach, and all escaped being dumped well offshore, in very cold water in gale-force winds, without lifebelts.

Another day ashore, Adam and I experienced a self-inflicted problem. Having climbed the highest point on the island, the 1,900-foot Mount Pond, we had great difficulty in descending through the numerous crevasses. In our inexperience we had failed to take into account the warming effect of the sun that by afternoon on a fine day had weakened the snow bridges over them, which we had crossed safely during our morning ascent, and we fell into one. Nevertheless despite our scare we managed to get out and completed a rewarding climb. Both incidents were a timely warning of a general principal not to venture out of one's sphere of competence without support and assistance from those who had the requisite specialised knowledge and experience in this dangerous environment. On the water we had our knives and we knew what to do in the case of the sinking scow, but in the crevasse we were ill-equipped, and struggled. I never climbed again unless in the company of an experienced Fid, and likewise made sure that Fids were always nursed in bad conditions when afloat.

From the top of Mount Pond we had a fine view of the Bransfield Strait, full of bergs and pack ice, the adjacent islands and, down in Whalers Bay, not only our own ship but also the *Oluf Sven* which lay nearby. She was supporting the Hunting Aerial Survey Expedition. The UK government had funded this expedition separately from Fids, with the remit of carrying out vertical air photography of the peninsula and off-lying islands with ground control from both their own and Fids surveyors. The Foreign Office had advised the Cabinet in 1954 that unless an aerial survey were undertaken, mapping the peninsula and underlining our territorial claim, we would be outflanked by the Argentineans attempting to reinforce their claim. They, it was thought, were preparing their own aerial survey of the area, possibly to be mounted with much greater ease from Patagonia. The Directorate of Colonial Surveys had awarded the contract to Huntings, and John Mott, a surveyor experienced in Greenland and the Himalayas, was appointed to lead a hurriedly put together expedition. Aboard the ship with John was Squadron Leader Saffrey, vastly experienced in aerial photography and reconnaissance, in command of the flying team. Two Canso flying boats had flown in from Canada, taking six weeks, each flying for 70 hours. The expedition also had two helicopters, all their fuel, a hut for erection ashore, steel for a slipway, and all the necessary living stores aboard the vessel. They were a very likeable, confident and professional group, and many of their ground staff had previously been with FIDS.

But their Danish ship, working the British coast when chartered, was a different matter. She was crewed by a ramshackle bunch of Scandinavians, and her master, though pleasant and reliable and I am sure a thoroughly competent seaman for normal commercial trade, had never sailed in either polar or uncharted waters. Not long before our meeting them, this combination of poor crew and inexperienced master had nearly lost them their vessel.

Returning to the extraordinarily more competent British seamen of the 1820s, Captain William Smith advised the senior naval officer in Valparaiso of his first sightings of the South Shetlands, but had been disbelieved. In consequence of reporting his subsequent landing upon them, he had his ship chartered for a voyage of verification and discovery.

Edward Bransfield, an RN master, was given command of Smith's fourth voyage to the area, in his brig the *Williams*. Sailing the length of the strait that now bears Bransfield's name, from close by Deception Island, on 30 January 1820, they sighted land to the southward. They turned towards their discovery of mainland Antarctica, and Bransfield became the first to survey any part of it, though modern translations of the logs of a voyage by Bellingshausen, just months earlier, may dispute this as being the first sighting of the continent. Bransfield called their find Trinity Land, after, as he put it, the Trinity Board, meaning the Board of the Corporation of Trinity House, the present-day lighthouse, pilotage and charitable body of England and Wales founded by Henry VIII in 1514. In Bransfield's time it was the pre-eminent marine charity looking after distressed seamen, their widows and dependants, of whom in those difficult days of sail there were many. Many years later, I was to become a member of that board. Smith, sadly, after returning home with little of commercial value, found his partners bankrupt, the charter monies lost and, despite his discoveries for which he unsuccessfully tried to obtain reward, became bankrupt himself and gave up deep sea voyaging. In 1824 he became a Trinity House pilot licensed for London and Harwich. He ended his days as a resident in the Trinity House almshouses at Mile End.

Towards the continent

Leaving Deception on a fine day we set course following the track of Smith and Bransfield towards then beside their Trinity Land. The faintly sunlit pastel-hued mountains low on the horizon some 80 miles distant, seem to beckon and invite a closer look. Perhaps it was the knowledge that this was the Antarctic mainland that we were seeing for our first time that really attracted us. The coast of this corner of the Antarctic is geographically the nearest to more temperate lands, and is therefore in one sense the most easily reached, but it also happens to be possibly the most unapproachable. Most of the continent has as its defence pack ice of a particular age or nature, which can be both assessed and worked through accordingly. Here the ice varies enormously within a few miles. It can also move at a fast rate of knots. Heavy Weddell Sea pack ice streaming around the top of the peninsula meets and becomes mixed with lighter, more local, ice. Blocks of shelf ice, rock-hard, the size of a house and larger, having calved from the Larsen Ice Shelf or the wide channels of semi-permanent ice of the eastern peninsula, are scattered amongst the pack for the unwary. The coast is extremely foul, and until much experience of its seemingly rather similar features has been gained, it is very difficult to position a ship from it.

In addition, the coast lies in the middle of the circumpolar trough of low pressure, 62–64 degrees south in this area, along which depressions travel. The weather in these latitudes is therefore extremely changeable, often severe, and most frequently westerly. The last factor makes the peninsula a lee shore, which in the context of sail meant that ships were likely to be pinned against that foul coast.

Our scant chart, with no more than the rough outlines of the mainland coast and islands, and the occasional sounding,[13] abounded with names from the past. Bransfield not only had his name given to the strait between the South Shetland Islands and the mainland, but quite rightly also to a nearby island and mountain. Captain Nat Palmer, possibly the second person to sight the mainland peninsula, had it named after him by the Americans; but in the 20th century international agreement was reached to call that peninsula – Bransfield's Trinity Land, the American Palmer Peninsula and our later Trinity Peninsula, all being the same stretch of land – the Antarctic Peninsula, overcoming the Trinity/Palmer Anglo-American disagreement. Palmer is now remembered not only by his name being applied to the large archipelago immediately to the westward of the mainland, which he did explore, but also to the southern end of the Antarctic Peninsula, still called Palmer Land – which seems incongruous, for he never went there and possibly never even sighted it. Astrolabe Island and Zelee Rocks are named after the two ships of Frenchman Durmont d'Urville during his 1838/40 expedition, and Mount d'Urville commemorates the man himself, whose descriptions of the southern continent are one of the most vivid. Ross Island, and Erebus and Terror Gulf, named after Sir

---

13 Depth of water.

James Clark Ross and his two ships, recall an explorer of both polar regions; he sailed south between 1839 and 1843, enduring appalling and frightening conditions, particularly when his two ships collided whilst trying to avoid being driven on to a mass of bergs. But neither d'Urville nor Ross lost a man, let alone a ship, on their southern voyages. Not so lucky were Nordenskjold, the Swedish explorer who led an expedition south between 1901 and 1904, and C. A. Larsen, the renowned whaling captain in command of their vessel the *Antarctic*; the east coast of the peninsula and its enormous ice shelf bear their names respectively. They lost their ship, crushed by ice in February 1903, some crew stranded and wintering on Paulet Island, relatively near the coast, and some at Hope Bay on the mainland, where we now maintained a base, but not to be visited on my first season.

One's thoughts cannot help but turn again to the exploits of those early mariners, with ropes and canvas frozen stiff, deck and spars coated and slippery with ice, hands numb with the cold, and the ships so small and their freeboard so low that they were often awash with icy water. Sounding was done by hand lead, and bearings were taken over an often wildly swinging compass with a primitive azimuth. They would have struggled to take sights from a heaving deck, through a heavily clouded sky, with unreliable chronometers and inaccurate logs and hourglasses to back up their sights, and they would also have had to deal with fog, which does not always come with calm, a common misconception. They were extraordinary; their adventurous spirit, courage, grit and determination are to be marvelled at and applauded. Since the time of my introduction to these waters, I have never lost my admiration for those seamen, and indeed as I grew more familiar over the years with this world of rocks, ice and storms, my respect grew immeasurably.

Turning southward down the Gerlache Strait, we then entered the waters lying to the westward of the Antarctic Peninsula. The coastline is heavily indented by extensive bays and inlets, and from close inshore to several miles offshore there are many archipelagos strewn with rocks and shoals. The peninsula itself maintains a mostly broad plateau of about 6,000 feet in height for its entire 600-mile length, with some isolated peaks rising above it to even greater heights. Significantly, though, its mountainous spine – the continuation of the Andean Chain and the Scotia Arc – lies on its western edge, close to the shore. Consequently these mountains and their heavily glaciated surrounds tumble within a very short distance from great heights to sea level with nigh-vertical rock faces, precipitously ending spurs and buttresses, steeply descending glaciers and ice sheets with chaotic ice falls, eventually fronting the water or hanging over sheer cliffs not far inland. As it enters the water, this terrain thus creates a coastline that is primarily of ice but often with a visible rock footing. Commonplace are headlands with 1–2,000-foot peaks shouldered by extremely steep ice slopes, flattening to an ice piedmont of only short breadth, abruptly ending with an ice cliff. Likewise islands with similar characteristics and proportions rise sharply out of the sea, creating narrow and deep fjord-like channels, with vertical ice or rock walls between themselves and the mainland. The major islands of Brabant and Anvers rise to heights as great as those of the mainland, yet are distanced only a few miles from it. Mount Parry on Brabant Island has an elevation of 8,274 feet, yet its shoreline is only three miles distant from the summit. Mount Français on Anvers

Island rises to 9,258 feet, one of the highest in the region. Its summit lies only eight miles from the channel, in places less than a mile wide, separating it from Wiencke Island, which itself builds in several places to peaks of 3–4,000 feet. Adelaide Island, separated at one point from the mainland by the Gullet, barely a mile in width, has four peaks of around 7,000 feet along its 70-mile length.

This scenario is excellent for navigation, except that the deep channels allow large icebergs to enter them, which can then eventually get caught, ground and block the waterway. Archipelagos of smaller islands, islets and rocks demand the opposite type of navigation, with foul ground and abundant grounded ice of a lesser nature. Ice builds on any rock face that is not near-vertical, and consequently snow-domes cap most of the smaller islands, rising gently on their windward sides but falling sharply into the sea with an ice cliff on their leeward. When working through such areas as the Argentine, Wauwermann, Biscoe and Dion Islands, it can be difficult to distinguish bergs from islands unless these two features of sloping and vertical extremities can be seen. But whatever surrounds the seascape, good anchorages are few and far between. There are some notable exceptions of sticky mud, such as Potters Cove on King George Island, found extraordinarily early by sealers who must have searched diligently; but the norm, as is to be expected, is a rock bottom. Many of the apparently sheltered bays and coves suffer from violent squalls and katabatic winds coming down off the higher ground. Sometimes semi-permanent floating ice shelves varying from six to fifty feet in height, to be found filling narrow channels, afforded a good berth for our ship. In just a few instances we were able to lie alongside sheer rock faces, but with both ice shelf and rock face berths, any swell often made this an uncomfortable or untenable option.

Shortly after commencement of this passage towards Port Lockroy I had my first chance of working pack ice alone on the bridge – but not for long! Most seasoned seamen live in fear and dread of ice, but in my experience almost all young men when tackling their first pack ice do so with far too much gusto. I was no different. My watch was the morning eight to twelve, and we had sailed just prior to that, Johnston taking the ship from the anchorage and out to sea. Come breakfast he handed over to me. That mixture of ice, referred to earlier, soon caught me out. The whole ship shuddered and rang with a spring-like resonance as we hit a particularly hard piece of glacial ice hidden within a softer floe, mounted it, slid off with an acute list to one side and then rolled violently back to the other. I removed the entire breakfast lay-up from the wardroom table. Johnston reappeared on the bridge, piece of toast in hand. 'I'll take over, Tom,' was reprimand enough; and another week passed before I was allowed my next attempt.

The navigational equipment originally fitted for the Baltic trade was very basic and of poor quality; gyrocompass and echo sounder were both often down. However, whenever possible on any passage we immediately began to survey. Save for the rare exception of harbours like Port Foster within Deception Island which had been surveyed properly during Operation Tabarin, and the occasional sketch survey and passage-making observations of earlier sealers and explorers, the coasts and surrounding waters were not surveyed. Furthermore, where an apparently safe line of soundings had been placed on a chart, the precipitous nature of the surrounding land mass resulted in similarly steep offshore pinnacles of rock often being missed just metres from it.

On every passage within sight of land we would prepare a plain Whatman drawing board and plot our position and course by dead reckoning down its centre. Whether our surroundings were distant or close to hand, we would shoot up[14] mountain peaks, glaciers, rock buttresses, headlands, islets, rocks and breaking shoals at those dead reckoning positions. Laying off those bearings from the positions plotted on the board as we progressed, we defined where they intersected the relative positions of those features. The radar, if working, helped us draw the coastline, though care had to be taken that the echo was coastal, and not grounded or fast ice[15] some distance offshore. Those with the ability – notably Adam, but not me – would also sketch the shoreline. At the same time as taking bearings, and continuously between, we would, when able to do so, take soundings along our track. Thus a running survey would be compiled. Over several passages, by taking a slightly different route, if possible parallel to that initially chosen, and at times somewhat bravely towards the shore, we would build up a workable chart. Stretching ourselves thus laterally was invaluable when on any later passage we were forced towards the coastal dangers by ice. Important, too, was that no scientific observations were worth much unless the observer knew where he was, and we enhanced our contribution to this by also occasionally taking astronomical sights by theodolite ashore at one of the features fixed on our running survey to position that point on the planet and thus locate our work.

On that first voyage we sailed the 90-mile length of the Gerlache Strait and Croker Passage, 15 miles wide at its northern end and narrowing to less than 2 miles navigable at its southern; mountains and polar plateau of the peninsula to the east, the majestic islands of the Palmer Archipelago, Liege, Brabant and Anvers to the west. Once again, names of those earlier explorers, or those chosen by them, surrounded us. The strait was now named after the Belgian explorer Adrian de Gerlache, although he had actually named it after his ship the *Belgica,* in which he had sailed these waters extensively. In 1898 he overwintered in her, becoming the first to do so in Antarctica. His first lieutenant, Roald Amundsen, who later beat Scott to the Pole, had a nearby mountain named after him. Dr Frederick Cook, also a member, wrote an account of this Belgian expedition *Through the First Antarctic Night*, and later, as did Peary, he made a dubious claim to have reached the North Pole. Baron de Gerlache allowed each of his men to name two features. The name of Brooklyn Island on the eastern side of the strait had initially surprised me, until I read in Cook's account that he chose that of his home town as one of his two names. We sailed through an almost invisible opening of apparently foul water, between Lion and Wiencke Islands, where land ice could not be distinguished from sea ice and rocks were hidden from view by the latter. Poor Wiencke, a 16-year-old deck boy, had been lost overboard from the *Belgica* not far from this location; perhaps of little consolation to him, he does have the most spectacular island named after him. Fourteen miles in length and narrow, it rises to some outstanding sheer peaks, and forms the eastern side of the Neumayer Channel. This was to where we then swung our ship, in loose pack ice.

---

14 Take bearings of.

15 Unbroken sea ice remaining attached to the shore.

What we then beheld was both astonishing and eye-watering; Norway, Spitzbergen, Greenland and Alaska have good combinations of mountains, glaciers and channels, but none match the grandeur of the major Antarctic archipelagos of this region. The mountains here are more precipitous and, as most rise from so close to the water's edge, so overpowering; nowhere else in the world are mountains and seascapes set in such juxtaposition. A problem in describing the scenery of the peninsula and its islands when sailing southwards is that it progressively becomes more spectacular. Very quickly all the adjectives have been utilised, and by the time this area is reached one feels incapable of doing justice to such grandeur and beauty. It was difficult to concentrate on the navigation for gazing at the scenery, made all the more exciting for there being no visible exit to the channel, it having a double dog-leg at its southern end.

As we rounded onto our course down the channel there came into sight an extravaganza of ice cliffs, vertical rock walls, mountain spurs with torrents of ice cascading between them, with tottering seracs and icefalls, all sat beneath the several thousand feet high sheer rock but snow-capped peaks of the mountains. A bonus in this magical place was that the myriad of colours within the ice had here a competitor; high up on the side of a 3,700-foot shoulder of Mount Français was a slab of lurid verdigris, naturally called Copper Peak.

Port Lockroy

Almost breathless we entered Port Lockroy, the rather grand name for a delightful but small cove named in 1904 by Charcot on his first expedition after the then French Minister of Marine. 'Boy Scout' Marr, who had sailed in the *Quest* with Shackleton on his final expedition, established the station here, at the southern end of Wiencke Island; Lieutenant James Marr, as he was in 1944 at the commencement of Operation Tabarin, chose the site after he had failed to land and erect a station at Hope Bay, an anchorage well known to sealers nearly a century prior to that.

A narrow entrance clear of dangers, with ice cliffs to port and low-lying rocky islets to starboard, opened out to a beautiful harbour surrounded by even higher ice cliffs than at the entrance, except for one rock spit upon which the base hut stood. Holding was reasonable, but strong winds frequently came down from the mountains and we would shift from the centre of the cove to lie close under the ice cliffs for shelter. The snow blew off them on these occasions and put us in our own private blizzard, but our position prevented us from dragging and facilitated the boat work. That was fairly straightforward, there being only a short run to the landing and with little fetch even in a strong wind for any chop to develop.

As we entered on that first occasion, our eyes were drawn to the ice monolith barely attached to the cliff face at its foot, and teetering at an angle, ready to fall. We commented to Johnston upon it, but he dismissed our observations, saying that it had been like that for years. As he turned to go down from the bridge, all hell let loose. The monolith calved; the tidal wave it created set off further icefalls, and in seconds the whole previously calm and empty bay was full of wildly gyrating bergy bits and brash. Eventually, while lying quietly to anchor again we trained our binoculars on the top of the totally snow-clad Mount Jabet just inland from the cove, for there a sailor from Charcot's expedition aboard the *Français* had driven in an oar, and it was rumoured that it was still there, but we saw nothing. Later, we attempted the climb to verify the story but failed. One sunny afternoon whilst we remained at anchor for our engineers to assist on the base, I did make a climb, with second mate Adam and a Fid called Wally Herbert, like us in his early twenties and on his first trip south, to some height up the slopes of nearby Doumer Island. They were both budding artists, whereas I was not. They set up their easels, set out their paraphernalia and worked in watercolours. My task as dogsbody was to add gin to their painting liquid, to keep it fluid in the low temperatures.

Adam certainly had an artistic lineage. His grandfather, S. J. Lamorna Birch, had been a prominent member of the Newlyn School of Art in the 1900s, and his mother, a talented artist, a friend and pupil of Dame Laura Knight. His father was also a writer on nautical subjects. Adam continually struggled with his painting but did achieve great success in hydrographic surveying, in which he was already showing great interest. After FIDS he went to Canada, where he became assistant hydrographer, and then to Monaco as a Director of the International Hydrographic Organization.

Meanwhile Wally, even then quite proficient as a photographer, never thought that he could paint well and only took it up seriously later in life. His oils of the polar world now sell for enormous sums, but his achievements went far beyond that. He became Britain's most accomplished polar traveller. He was knighted at the time of the millennium honours, having had his triumph, in May 1969, of successfully completing the first unsupported Arctic crossing, totally overshadowed at the time by the simultaneous first lunar landing. Taking 16 months, with dogs and sledges, overwintering on the pack ice, and transiting both the Pole of Inaccessibility and the North Geographic Pole, his feat, in the words of the expedition's patron, HRH Prince Philip, 'ranks among the greatest triumphs of human skill and endurance'. Wally loved and understood the polar regions and the people of the polar north, spending a total of 15 years entirely in that world. He lived and hunted with the eskimos for two of those years, and sledged in total in excess of 25,000 miles. I am very fortunate to have had Adam as a friend from then until now, and very privileged to have known Wally from such an early age until his untimely death in 2003.

One of the bonuses of serving on the Survey's ships was that of working alongside expeditioners with such an extensive variety of disciplines, and whose enthusiasm for their subjects, whether they were scientists or surveyors, meteorologists, mountaineers, diesel mechanics or builders, was unbounded. Further, they all, from a combination of youthful exuberance and being volunteers in a common exciting adventure in a magical world, always had a desire to share their knowledge.

I went ashore for another 'jolly' at this base with a young glaciologist. We landed from our launch and crawled onto the slabs of rock beneath a section of the ice cliffs, venturing as far as it was possible to go. He showed me the gouges in the underside of the ice corresponding with the rock features it had passed over, as evidence that it was slowly but constantly moving seaward, grinding over the rock base it sat on. That first fairly simple but enjoyable science lesson was made all the more exciting for me by thinking about that tremendous collapse of ice that we had witnessed on arrival in the anchorage.

Danco Island

We had tasks other than relieving the bases that season; our main duty was to site and build two new stations that we carried. This task required much exploratory navigation of the coast in totally uncharted waters, although many of the channels and bays we entered had seen sealers, whalers or expeditions in them briefly at some time. Our remit was to find sites for the new bases from which the Fids could get to the peninsular plateau to progress their topographical survey, establish triangulation points for the control of the forthcoming aerial survey and carry out geology. In the winter months it was also planned that they would travel around the coast on the sea ice, visiting nearby islands to the same ends. The master, knowing these objectives, had total discretion as to the placing of the stations. We searched extensively for a suitable site, steaming in and around inlets and bays over a 20-mile stretch of the mainland coast as well as exploring the nooks and crannies of Brabant and Liege Islands. It became apparent that it was going to be immensely difficult to find any area of bare rock that was level enough and of sufficient size to set out a base. To combine that with a safe anchorage to lie at whilst building, and for it to be judged accessible again during any following seasons of more extensive sea ice, was nigh impossible.

Then HMS *Protector*, the Falkland Islands guard ship, made one of her short appearances, standing off in the Gerlache Strait. Rendezvousing with her, Adam took a flight in her helicopter to make a further reconnoitre from aloft. He returned with the news that there was nowhere better than a site we had seen already, on Danco Island in the Errera Channel on the eastern side of the Gerlache Strait; this was close to where we had earlier landed some Fids on the mainland to explore the possibility of finding a route to the plateau. Returning there and reunited with the climbing party, we were to learn that it was most unlikely for any such route to exist, but that on Danco Island there was a good site to build a base. With difficulty we also found an anchorage, and paused whilst messages were exchanged with HQ as to its suitability and whether or not to build there. The reply came in the affirmative.

But Johnston was later criticised for his choice of location; Dr Fuchs, later Sir Vivian, in his history of FIDS, *Of Ice and Men*, claimed that Johnston was more interested in a good anchorage than an appropriate base site! Fuchs was an outstandingly good leader and boss, and this criticism by him of any of the seafaring staff was very unusual. He was normally very complimentary about us, saying that our efforts in support of the survey were invaluable, yet mostly went unsung. At the time he was on secondment to lead the Commonwealth Trans-Antarctic Expedition and reconnoitring for his base camp at the foot of the Weddell Sea in the *Theron*. The governor, whilst personally keen on spreading FIDS research activities, had his masters in the Colonial Office; they had very much in mind the establishment of a well-spread occupancy of the region to display sovereignty. We had searched extensively for sites to suit the objectives of the survey, but having found none, and indicated so, pressure was put on us to build somewhere, almost anywhere, with the materials we had aboard before the season expired. All Fuchs' information on the subject was related to him later by those behind desks in London who steered the blame away from themselves and onto Johnston. Both Adam's journal and my diaries note that we paused a full day before commencing to build, awaiting a reply to Johnston's signal elaborating the drawbacks of the site. The reply came back: 'build'. But regrettably this base and that of Detaille Islet, which was also established hurriedly that season by the *John Biscoe*, both failed to provide access to the plateau and never had sea ice sufficiently stable to travel upon safely in winter, and were consequently closed after just two seasons.

Danco Island, which gave its name to this new base, and the adjacent mainland of Danco Coast were both named by de Gerlache in memory of Lieutenant Emile Danco, who had died during the 1897–98 Belgian Expedition; a death brought about by his having a weak heart, exacerbated by his unwillingness to eat freshly caught seals and penguins to obviate scurvy. The anchorage was far from satisfactory, not one that we would normally have chosen to lie at for any length of time; the bottom was very uneven, with sharp pinnacles and deep trenches into which the anchor slid, and the tidal stream through the channel often ran at 3 knots, carrying with it all manner and size of ice. We developed a practice of paying out the anchor cable to its extremity as a particularly dangerous berg approached and then, when it had drifted very close, heaving in again, if necessary applying a touch of engine power but often just by using the helm, thus canting the vessel in the tidal stream and walking ourselves past the berg as quickly as possible, hopefully but not always, without hitting it. With a wind of sufficient strength and a direction to lay us across the current, bergs under way in it would come at us on the beam. Frequently they would then come to rest alongside. There was only so much cable work one could manage whilst working the boats alongside, so taking great care of them and having only a limited season in which to accomplish our tasks, our expedition mentality led us to accept much punishment; our well-indented hull bore evidence of this attitude by the end of the season.

Despite the difficulties of lying off Danco Island we enjoyed our visit there. The island was only 600 feet in height and almost totally snow-clad, the exception being a small coastal strip of rock and shingle rising gently from a beach. This was home to a few thousand penguins. Matters of conservation were not as high a priority then as they are now, and in order for the base to be built we had to ask several hundred of them to move! The island had a good ski slope, but again we had to compete with the penguins. For no other reason that we could fathom other than for the sheer pleasure of it, they would climb to the summit, waddling up in line ahead, pause momentarily, and then toboggan down on their fronts, steering with their flippers, the whole way back to the shore. This they did repeatedly. Our slalom technique improved immeasurably avoiding the amusing little creatures.

The base was built successfully on short concrete pillars, with restraining wires from ground anchoring points over its roof. By the time we departed, generators and a radio office were installed and working, although the hut was not quite ready for occupation and the base members were left housed in tents on the beach. But once the base was operational, no sledge access to the plateau could be found, and the local sea ice, probably because of the strong currents offshore, never formed sufficiently well to be relied upon for travel.

The latter problem led to a period when the Fids here undertook much boat work to achieve their aims. This also became the practice at other bases to further their work, all with the cognisance of HQ, but we mariners were highly critical of this. That season we had already been alerted to a search for some Argentineans lost in a dinghy from Hope Bay, and had heard of a similar scare from one of our own bases. Not only were there the dangers of bad weather and collision with ice, their outboard motor propellers being particularly vulnerable, but it was not unknown for leopard seal to attack the occupants of small craft.

Anvers Island

Leaving Danco Island at the end of February we steamed to the Anvers Island base, established only the previous season to survey topographically and investigate the geology of the island. The base here had a unique inception. A sample of rock from the Anvers shoreline of the Neumayer Channel containing nearly 50 per cent copper ore had been brought into the London office, and on that evidence of minerals possibly being found there, together with the desire to proliferate stations in the area, plans were made to explore the island. The Norwegian vessel *Norsel* was chartered by FIDS for the 1954–55 season, and established a base for just six men to carry out the survey. The island was ranged over extensively and the first ascent of the near 10,000-foot Mount Français made, but no route to the most likely area of copper deposits, near Copper Peak, was found. However, the area was then reached with the assistance of the Fids 40 miles away at Port Lockroy, who left their base by boat, crossed the Neumayer Channel to Anvers Island, and then shifted the Anvers survey party along that island's coast, but no indication of any copper deposits was ever found.

It was a smart, well-organised and well-maintained base. Adjacent to a small, pretty, almost entirely ice cliff-encircled cove, but with ample danger-free swinging room, it had a very difficult rock-strewn approach which was to test us severely on arrival. We had no time to complete any soundings but on an island that would lie ahead on approach we did erect a rock cairn to steer on and to keep on a bearing astern when leaving. Even then, having entered successfully, if we were pushed only a fraction off-line by ice when departing, the proximity of shoals to port and starboard was far too close for comfort.

The Argentine Islands

From Anvers we headed south towards the Lemaire Channel, which must rank as one of the most spectacular mountain channels in the world. We held a steady course with Cape Renard right ahead, its twin 2,000-foot sharp peaks of incredible beauty piercing the sky. As the channel opened to our view, we altered course to pass close past those dramatic peaks rising steeply out of the water, and then into the narrow waterway with sheer rock and ice walls rising overpoweringly on both sides. So vertically did they rise for its entire ten-mile length that hardly a foothold was possible from the water. Though the channel was deep, we were often forced to within feet of the side walls by bergs or large pieces of pack lying midstream, then into centre channel again on meeting more ice aground, deep beneath the surface against the invisible narrowing rock sides of the underwater chasm. Here we encountered other floating ice under way at speed either in the fast-flowing current of centre channel or travelling in the opposite direction in a strong counter-current between the main centre stream and the sheer sidewalls. The combination of such unbelievable beauty and such exciting navigation was hard to comprehend, but within a couple of hours we were through, and the Argentine Islands, our next destination, now lay ahead, some ten more miles down the Penola Strait. This group of low snow-clad islands and islets with numerous small boat channels between them is the nearest one could imagine to a polar paradise. Unlike the scenery of the Lemaire Channel, whose mighty grandeur could only be gazed upon, these islands, whilst set close to an abundance of similar high splendour, were possible to travel over and in between, to physically enjoy.

Having hugged the ice cliff to starboard to enter Meek Channel between two of the main islands but also to avoid Corner Rock, situated in midstream, we passed through the 200-foot-wide entrance and executed a running moor off the base to lie tight to both anchors, for there was very little room to swing in. The 'old' *John Biscoe* arrived and strapped alongside, facilitating the transfer of people and cargo. She certainly looked the part. Despite the modern and quite pretty appearance of our frail and underpowered steel vessel – though neither of these latter characteristics were obvious or were known to them as yet – she and her crew made us newcomers feel not quite in the same expedition league as them and their wooden ship. Built in America as a boom defence vessel, of pitch pine planking on oak frames and with diagonal green heart sheathing as protection against ice, she displaced 1,015 tons and had a length of 194 feet, a beam of 34 feet and a loaded draught of 14 feet. Her twin diesels powered an electric motor providing 1,200 horsepower with a best service speed of 12 knots. Her small numbers of Fids were housed in two bunkrooms, the hammocks often slung over a sloshing ingress of water. She had the reputation of being an uncomfortable ship in a seaway but excellent in ice, when greenheart planks were seen to be torn from her hull and rise to the surface as she worked it. Her master, Norman Brown, had been chief officer under Captain Johnston on her previous voyage.

Our duties lay in our work with the base and our sister ship, but we could not help being distracted to gaze in awe at the surrounding scenery. To the north lay the ice-clad peaks and glaciers adjacent to the Lemaire Channel, which we had passed through; across the ice-choked Penola Strait, five miles away but appearing much closer, the mainland rose for the extent of our vision from north to south, to several substantial peaks of 6,000 feet, Mounts Scott and Peary amongst them, with the striking sharp and isolated Cape Tuxen rising to 3,000 feet in the foreground. The Argentine Island group itself was intriguing, and invited exploration, with innumerable narrow but tortuous passages between the islands, which had easily accessed ice foots to shoreward of gentle snow slopes facing the prevailing wind, but precipitous ice cliffs to leeward. One such passage, Stella Creek, led past the place where the 1934 British Grahamland Expedition had wintered its vessel, *Penola*, and onto Wordie Hut, its base. Some years later this hut completely disappeared, thought to have been washed away when one of those high ice cliffs calved and sent a tidal wave though the adjacent narrow creek and over the low-lying site. We worked our cargo all too quickly, being so close to the base and having *Biscoe* alongside, but with our season being so short we had to move on. Fate had other ideas, though. We lost one anchor and several shackles[16] of cable. Finding it from the workboat with a grappling iron, but not being able to lift it, we buoyed it off. Then by manoeuvring the ship over it, we were able to bring some cable to the surface with the crane, but we had snagged a length near the anchor. Two days later, by hanging off the anchor, breaking the cable and reversing it, we put all back in its proper place.

Our next anchorage was just for an afternoon off Peterman Island, only a few miles distant. There we went ashore to replace a plaque which had been found lying loose in the snow near the small summit of the island by some thoughtful soul, and had been sent back to France. It commemorated the stay of Dr Charcot in 1909, when he had wintered his vessel in a well-chosen inlet; he had strung a chain across its entrance, preventing any large ice moving in from the strait and damaging the ship before fast ice formed to protect it. The French authorities had asked FIDS to return the plaque to its proper abode! This we did, cementing it to a cairn we had erected, taking photographs as evidence.

Furthering the initial discoveries by Smith, Palmer and Biscoe, amongst others, de Gerlache and Charcot carried out the next stage of exploration in this region, charting and understanding the nature of it, as much and as far as weather and ice permitted, in their small vessels during the first decade of the 20th century. Jean Baptiste Charcot, the son of a wealthy French doctor, spent his inheritance on his two polar expeditions, and generated enough support and financial assistance from the French government and academies for them to become French national expeditions. He built two ships; the 105-foot *Français,* a three-masted topsail schooner of great strength but of poor speed whether under sail or power, for his 1903–05 expedition, and then the *Pourquoi Pas?*, a beautiful three-masted barque of 131 feet in length, also very heavily built, for his 1909–10 expedition. Both ships overwintered, and both were badly damaged more than once when they hit rocks during their exploratory passages. Both expeditions also suffered incidents of extreme danger and times of severe hardship. It was from here at Peterman Island that Charcot most probably came nearest to losing his life in the Antarctic. Crossing the five-mile width of Penola Strait to reach the Berthelot Islands and mainland, with two companions on a supposedly short excursion by small boat, their return journey was blocked by an influx of pack ice. The *Pourquoi Pas?* eventually searched for the trio after they had failed to return to Peterman Island, and only just in time saved them, as they were running out of provisions and were ill-equipped for a lengthy stay – but then she went aground herself, being badly damaged in the process. After wintering in the cove above which we had replaced his plaque bearing all the names of his expedition members, Charcot reached 70 degrees south making the discovery of Charcot Island, which he named after his father! His ability as a seafarer, and enthusiasm for exploration knew few bounds. He continued such that not long after these expeditions he served during the First World War, both in the Royal Navy, being given command of an auxiliary cruiser with the rank of captain, during which time he was awarded the DSO, and then in the French Navy. For several years after that, he again commanded the *Pourquoi Pas?* on voyages to the Arctic, but in 1936, when he was aged 69, she was driven ashore in a storm off Iceland and was lost along with all but one of his crew.

---

16 A shackle of cable is 15 fathoms, i.e. 90 feet.

Horseshoe Island

Only on the next passage southward towards Horseshoe Island did we cross the Antarctic Circle, 66.5 degrees south, such is the extent of the southern polar continent. Passing to the west of Adelaide Island, we crossed this circle of latitude marking the point where the sun remains above the horizon for one whole 24- hour period in midsummer, and conversely remains below it for the same period in midwinter. Between it and the South Pole these periods of total light or darkness increase such that at the South Pole there are six months of each. Whilst we enjoyed near 24-hour daylight in the mid-season, the darkness earlier and later on made passage-making much more hazardous.

Reaching the western extremity of Marguerite Bay, named by Charcot after his wife, we found that Horseshoe Island, where our base lay, had a very steep-to shoreline. After the *Norsel* had built the base on Anvers Island the previous season, she had sailed south to seek a site on the periphery of Marguerite Bay to replace Stonington Island station, 50 miles further south; Stonington had proved too often to be inaccessible to the ships for relief on account of extensive fast ice, and was to be closed. The *Norsel* had found no anchorage at Horseshoe in sensible depths, but had managed to secure alongside a vertical rock face close to the base site. We too found no proper anchorage, and feared not only poor holding but also being hit by the many large bergs under way in the vicinity. We therefore nosed right up to the shore, but found it hard to find a sufficient length of flat-sided cliff to lie against. Eventually compromising, we did manage, and with lines ashore belayed around rocks, we commenced to discharge as if in port. But the deep water still held the same problems; sizeable bergs would frequently drift alongside us. With no swell they were not a problem, but with even the slightest they became so. We could do little about them except use poles to encourage them to travel a bit faster in the direction that they were heading, and hang baulks of timber over side between the hull and any rock projection, to soften any shoreside effect of their impact.

One of our entertainments when worked had finished for the day was the showing of a film, it doing the rounds of Wardroom Fiddery, as the expeditioners mess was called, and crew's mess, during a week. We employed a mixture of naval and merchant service terms and behaviour, reflecting the mix of our complement. As we changed to the last reel of our film in the wardroom, the projector fell to the deck as a large berg collided with us, and remained alongside repeatedly nudging. Johnston announced in his phlegmatic manner that we should see the last short reel, then deal with the berg! But on this occasion we were forced to move off our berth, and let it pass clear before returning, because it was threatening the bridge wing and lifeboats.

Horseshoe Island had several significant peaks over 1,700 feet; one of these is now named Mount Searle after Derek Searle, the surveyor on base at the time of our visit and base leader the following year; both he and his wife Petra were to become prominent members of the Directorate of Overseas Surveys. However, at that time the Directorate of Colonial Surveys, established in 1946, was responsible for all Colonial and Protectorate Topographical and Aerial Surveys; shortly after our visit to Horseshoe Island it changed its name to the Directorate of

Overseas Surveys, becoming an Agency of the Overseas Development Agency, and drawing up maps to be printed by the Ordnance Survey.

Mount Searle, previously named Gendarme Peak, had its name changed in pursuance of the policy of the Antarctic Place Names Committee to keep historically associated names of features, and to change names of no particular significance to names with historic connections. This Foreign Office Committee now comprises, together with experts on the region, members from the government for the British Antarctic Territory, the British Antarctic Survey, the Royal Geographical Society, the Scott Polar Research Institute and the UK Hydrographic Office, who are responsible for recommending names within the Territory and South Georgia to be formally approved by the commissioner. I could understand the changing of names such as Gendarme Peak to Mount Searle, for the former was just an idea of a gendarme standing guard over the island – though I am not so sure of the wisdom of changing the admirably descriptive names such as Pin Point on Livingstone Island to Renier Point, when the headland in question actually did, from one particular direction, look exactly like a pin.

I climbed Gendarme Peak with two experienced mountaineers from the base. They were general assistants, the term used for those with supportive skills to run a base, but some also, as these men were, chosen for being proficient skiers or climbers, and used to setting up and maintaining a campsite, and who could turn their hands to sledging and leading the field parties of surveyors and scientists. Our approach to such climbs had to be extremely cautious on account of the severe terrain and weather, and because no risk to life or limb could be taken in leisure pursuits. Not so that approach for the Governor of the Falkland Islands, Sir Raynor Arthur, who had joined us from the *John Biscoe* to carry out an inspection of his territories. He was evidently a skier of some repute; an Olympic trial had been talked of. He arranged a downhill course on the steep slopes of Gendarme Peak, and after a couple of days of practice, we held a time trial. The base Fids were certainly experienced, some to a very good degree. Adam and I and a few others from the ship were novices, but we were budding surveyors and could record times and measure distances. Speeds of 80 m.p.h. were recorded as we rattled down the glare ice with barely any snow cover; we were either poorer surveyors or better skiers than we thought! The governor, of course, won.

Before leaving Horseshoe Island an incident temporarily halted our progression. We had only one doctor aboard this voyage, 26-year-old Malcolm Evans, on his way to winter on the island. Our donkeyman, Mackenzie, the senior hand in the engine room, developed appendicitis, and Dr Evans decided to operate. Although Johnston found as quiet an anchorage as possible, tucked between islands and in ice, unfortunately some swell still found its way to where we lay, but the gentle movement did not deter our young surgeon. After breakfast we cleaned the wardroom table and prepared well around it. Adam and I became theatre assistants; he having the task of keeping the patient anaesthetised with ether pads whilst I passed the requisite instruments and held a bucket for any disposables. Despite the slight movement, and vibration that we could not obviate because we needed to maintain the power from the generators, the operation went according to plan. The patient recovered well and was up and about on light duties a few days later – but Malcolm Evans was severely reprimanded by Dr Slessor, the chief medical officer in Stanley, for acting too hastily, implying that the patient should have been brought back to Stanley for the operation. Stewart Slessor, a kindly Scot, was an RNVR lieutenant and ex-Fid, having served at Stonington Island with Fuchs in 1948. I thought his criticism of Evans very unfair. To have returned the patient to Stanley Hospital would have meant breaking out of the ice to seaward and then undertaking at least a six-day passage northward, mainly across the lively Drake Passage. Once such a passage had been commenced there would have been no possibility of operating if the need arose. On reflection, I suspect that Evans did not even try to speak to Slessor before he operated, and that was his crime.

Stonington Island

From Horseshoe Island we ventured down the mainland coast to Stonington Island. Originally established by the US Navy Services Expedition under Admiral Byrd between 1939 and 1941, it was named by the Americans after the great home port of sealers and whalers on the US eastern seaboard, Captain Nathaniel Palmer being one of its sons. A hut was built there by FIDS alongside that of the Americans in 1946, but then abandoned in 1950. The approach was spectacular; as we rounded the high bulk of Millerand Island, Neny Fjord hove into sight with its imposing glacier, whilst the orographic clouds, lifting over the long high ridge of the peninsula, poured down the mountainsides, disintegrating as they dropped, bathing the coastal panorama in brilliant sunshine.

The station, however, was sad; it was dilapidated and had some four feet of ice and snow within, which had entered despite the care obviously taken to board up doors, windows and chimney. Outside, remnants of stores and wooden crates, coal sacks and empty oil drums lay about. Where seal carcasses had lain in readiness as food for the huskies, skin, blubber and other detriment remained stuck to the rocks, putrefying where it became exposed as the summer sunshine eroded the snow cover. It stank. What made the site even more depressing was the contrast between its miserable state and its gorgeous setting, and memories of its significant British and American heritage. Surrounded by mountain peaks and snow slopes, adjacent to a beautiful, wildly fragmented glacier front that continually calved into the sea, it was to here that the several hundred-mile pioneering sledge journey came in 1948 from Hope Bay, at the very northern tip of the peninsula, and from which many pioneering exploratory sledge runs were made inland and southward. Teams of ten dogs each would often have been spanned out nearby, ready to be hitched to their sledges for the off. This was where Fuchs cut his teeth in polar travel and logistics. Finne Ronne, the Norwegian-American leader of his own expedition, utilising the huts of the previous American one, wintered with an aircraft, making incursions over much virgin land, and made some outstanding sledge journeys between 1946 and 1948. Ronne had with him his wife, Edith, and his deputy, Darlington, had his wife, Jenny, with him as well; they were the first women ever to overwinter on the continent. We left the island vowing that if it were within our power we would return one day and clean the place up. In fact, we did so in 1957 when the base was reopened.

Ambitiously we made our way south from Stonington towards Alexander and Charcot Islands to seek what would have been our most southern base site from which to survey those areas. It was not to be. We encountered too much heavy ice to approach either. Few ships had ventured this far south down the peninsula, and not only was it totally unsurveyed, but no identifiable pattern of seabed geography could be established. There seemed to be shoals where reasonably deep water was to be expected, and vice versa. We got inside one such set of reefs and took several hours twisting and turning, unable to retrace our path of entry, before extricating ourselves. There being just too much ice and too many shoals, and the season nearing its end, to be caught so far south in late March could have seen us suffering the same fate as de Gerlache in the *Belgica* in 1898 in these whereabouts, becoming beset and overwintering; so we turned north. Weight was also added to the argument for departure when we suffered a long engine breakdown off a newly discovered group of major islands holding the pack ice together where we had anticipated good water. We passed the Faure Islands, where in a fierce gale Charcot had been driven out of control through rocks and reefs in the *Pourquoi Pas?* Grounding once in the gale, the ship had miraculously escaped total disaster.

Northwards once more

We entered the Laubeuf Fjord, attempting the inside passage – the half-mile-wide Gullet between the 80-mile length of Adelaide Island to port and the lofty peaks of the mainland to starboard – only to be thwarted by unbroken fast ice. We were thus forced to make an almost complete circumnavigation of Adelaide Island, and by passing through the Matha Strait, regain the calmer inside passages that would lead us past the Argentine Islands to Port Lockroy. Negotiating this glamorous but engaging route, part of it within the Grandidier Channel, we met the *Olaf Sven* once again, landing surveyors by helicopter to establish triangulation points for control of the aerial survey. We sighted her towards the western side of the channel, where she had been following a single sporadic line of soundings which came to an abrupt end on an otherwise empty chart save for some coastal outlines. They had probably been put there by an early explorer, stopped from going any further by ice or a shoal. The *Olaf Sven* had also stopped abruptly, and we became aware of her amongst the sea ice against a backdrop of the ice-domed islands when she used her Morse lamp to flash SOS. We quickly made radio contact with her and were relieved to hear that she was neither aground nor damaged by ice. She just wanted advice: 'Where do we go now?' was her extraordinary question. We felt like replying rudely, but instead closed her sufficiently for our motorboat to bring her master, Captain Ryge, aboard. Johnston then, in a kindly manner for him, gave him a one-hour lesson on working in uncharted waters.

In those days, neither vessel had side or forward scan sonar. Our sounding machines recorded on wet paper that often partially or completely froze when the bridge doors were open. The coldness and density of the water made for very poor echoes, and ice would get under the hull preventing the transponder from working properly. So, much of the time when within the 100-fathom line, we resorted to walking the anchor back one shackle, and when it touched bottom, we at least had warning of it rising. Also, we used our eyes. Tidemarks on grounded ice were a highly useful indicator of the whereabouts and depths over shoals. At any time but high water, a piece of ice grounded for a period would show an obvious area of smooth erosion below the high water line. At low water, beautifully sculptured mushroom shaped bergy bits were to be seen where waves had worn away the ice at the changing water levels of the rising and falling tide around its base, leaving untouched a much larger head. We were able to judge the size of such grounded ice and the depth of water in which they might be aground, and compared that with our own draught. Eddies caused by the rising bottom configuration were often spotted, especially when enhanced by any small fragments of ice that became caught in the excessive movement of the water rather than remaining wind-driven. The lie of the land when close was initially our best guide to offshore shoals. Low promontories always gave a warning of shallower water offshore, as did a series of peaks inland aligned such that they pointed seawards, when invariably further underwater peaks and breaking shoals were found to be on that same line and at the same intervals as between the peaks ashore. Glacial bays usually provided danger-free water, but any moraine debris was not always in line with the glacial edge. Heart-stoppingly but ultimately amusingly, we would sometimes thunder astern for a piece of black moraine adrift and embedded within floating ice, looking like a solid rock rising from the depths. A dark seal motionless between floes often gave the same false impression when we were apprehensively working shoal water. It was highly educational for me to listen to Johnston briefing Ryge, for he never taught us directly; we learnt mainly from observing him. I learnt the basics of such exploratory navigation on this first voyage, which I was able to develop later.

Making our way north we called at all bases en route, lifting outward mail, topping up their fuel for the last time and embarking any Fids who were leaving their southern home after, usually, a two-winter tour, resulting in a period away from the UK of nearly three years. Once he began his tour of the Dependencies, Governor Arthur had become obsessed with refuting the Chilean and Argentinean territorial claims, which were escalating. Wherever possible during our passage north with him we visited their bases, sought out any of their refuge huts or tented encampments and delivered Notes of Protest to their occupants, informing them that they were trespassing. They in turn, would furnish us with a similar note from Santiago or Buenos Aires. When the governor came ashore with us it was a fairly explosive affair; our two antagonists' habit of putting up boards in their national colours with an inscription as to their claim so infuriated Arthur that it led him, armed with a sledgehammer to vent his fury at such a violation of British sovereignty by smashing their boards to pieces. When left alone to do his bidding, however, we would deliver our protest on behalf of Her Majesty and then in turn amicably receive theirs, exchange some wine for whisky and, given time, share a tot. Then we would remove their boards, replacing them with our own Union Jack-painted masterpieces made by the ship's carpenter, declaring British Crown Land.

At Port Lockroy we met up with a 'detached' naval survey party, from their parent vessel, HMS *Protector,* led by Lieutenant. John Wynne Edwards, comprising several men and a launch. They had been put in to survey a safe anchorage for the Royal Yacht *Britannia* that was to attempt to visit the area the following season. Prince Philip was to undertake a world tour in her, and after attending the Commonwealth Games in Melbourne and thence to New Zealand, it would be possible in a season of light midsummer ice for her to call at the very western extremity of the peninsula on her way to the Falklands. The seaward approach to Anvers Island, at whose southern end we had that particularly attractive station, was through the Bismarck Strait, a deep, wide stretch of water clear of dangers. Mount Français on Anvers Island, one of the highest and the most westerly mountains in the area, should be easily found and identified, even if it meant waiting for a break in the weather in the event of no astro-sights being obtained on the final days of her passage. It was an ambitious but good plan; only the ice situation of the coming season was unknown. John surveyed a clear patch of water of appropriate depth easily accessed from the strait. Task completed, with the unbounded enthusiasm of all naval hydrographic surveyors that I met, he undertook a circumnavigation of Anvers Island in his small launch, after which he encouraged us to take him to new pastures. We returned with the ship to the northern end of the Grandidier Channel, south of the Argentine Islands, from which, with our assistance, he surveyed as much of that channel as time allowed. It was invaluable experience for us; John was as good a tutor as he was a surveyor. When the naval party departed, the *Shackleton* lay off and worked the new base at Ferrin Head. Johnston let Adam and me carry out some further survey work amongst the nearby Fish Islets in one of the ship's lifeboats, the motorboat being occupied with runs to the base. Unfortunately we holed the aluminium hull of the lifeboat in ice after a few days, returning to the ship like the Dutch boy with his finger in the dyke, pumping and baling furiously, which brought that enterprise to an abrupt conclusion. There were, however, very strict parameters to the season dictated by the onset of winter, which meant that by then, well into April, regardless of that incident our time south was coming to an end.

As temperatures drop, the seawater between floes freezes, firstly in the southern areas of the Antarctic seas and very slightly later further north. This limits any pack ice movement to that of only expansion, the newly formed ice between floes preventing the older floes coming together again. This process of expanding ice, by which any open water revealed is then frozen, makes for a continual northward march of pack ice in autumn. We were most affected by this occurring in the Weddell Sea, from which pack ice begins to stream up towards the South Orkney Islands, then around the tip of the peninsula into the Bransfield Strait up against the southern shores of the South Shetland Islands, and then much is borne down against the west coast of the peninsula, filling that sea area. With far fewer hours of daylight, worsening weather and temperatures dropping, and with the fear of becoming entrapped in ice possibly for the winter, it was time for us to leave. We left the pristine shores of our polar world for the grey tumult of the Drake Passage – but not before one last encounter with pack ice – and the Argentineans and Chileans and their offensive boards.

Homeward via Stanley

Once back in Port Stanley, we had a few days' respite before the voyage home during which we changed the ship's Articles of Agreement. These were the standard merchant service terms of agreement between the master and his crew, opened at the outset of each voyage. Although we had commenced this first trip in Southampton, the ship was registered in the Falklands and that was to be where we would normally commence our round voyages. On this visit and at this stage of all later voyages, we signed off the crew, most being re-engaged immediately. Some, on that initial voyage from Southampton, chose to remain in the Falklands and were replaced by Falkland Islanders. The Fids also signed on as supernumeraries, with a rate of pay of one shilling per month, which some asked for but never got! Many disliked the idea of putting themselves under the authority of the master, but we needed them to be under our orders to work the ship, and she not being classed to carry passengers, there was no alternative if they were to travel with us. Some Fids also occasionally remained in Stanley after their tours south to take up positions ashore with government, the Falkland Islands Company, or indeed to work in our own FIDS office.

One of our young seamen, John Smith from Southampton, on a slightly later voyage after over two years with us went ashore, married an Islander and joined the Falkland Islands Company. He stayed there for 20 years, then resigned to follow his own pursuits, having built up a great knowledge of Falkland Islands history, particularly maritime affairs. He became an accomplished artist and author, served on both the executive and legislative councils, was instrumental in founding the Falkland Islands Museum, becoming its curator, and also became involved in the formation of the South Georgia Museum and the SS *Great Britain* Project. His further achievements would fill the page, but lastly it must be said that he played an important role during and after the Falklands War for which he received the MBE, and his book describing the islanders' experiences under the Argentinean occupation, *Seventy-Four Days*, is both poignant and humorous.

His leaving us, along with those others who chose to do so throughout our voyages, created a system of exchange whereby Falkland Islanders were able to join us. Despite being islanders, they were not, in the main, natural seamen; very little fishing was undertaken there, and only a few very small inter-island vessels and the one Falkland Islands Company small passenger/mail ship plying between Montevideo and Stanley and then around the islands, provided jobs at sea. Over the years we employed and trained many from the islands such that at times our crews were mostly from the Falklands. As a result of these exchanges and our annual return to Southampton, where many Falkland Islanders chose to remain, a large Falkland Island community developed there. More importantly for us all, though, was the unique relationship which grew between our two ships and all the Falkland Islanders. Those who sailed with us were tough, likeable and reliable; I came to appreciate their good humour and companionship, their dedication to the task in hand, and their ability to work long and hard when conditions permitted and yet lie up content when they did not, which served us so well south. It also led me to tolerate their inclination towards a drink when off duty ashore in Stanley, which often affected their abilities when departing; with the mates occupied fore and aft it was not unknown for me to man the bridge on my own on such occasions! Lastly and most importantly, I appreciated their loyalty that at times I must, once in command with my autocratic style, have tested severely.

Standing by refit and leave

We arrived back in Southampton after our five-month voyage on 31 May 1956 to a rapturous welcome from relatives and friends of both crew and Fids. There was local press and television coverage that made much of the obvious scrapes and dents from our encounters with ice and which applauded the adventurous nature of our voyage in their reports. From our arrival berth, No. 37, at the very outer extremity of the docks, we firstly went to Thornycroft's shipyard at Northam up the River Itchen, where we stripped the ship bare of stores to commence the refit and thence into dry dock. There we found that we had sustained extensive damage to the hull below the waterline. Mostly forward, where one would expect, but also heavy corrugations to the bottom plating about one third of the ship's length from forward. Such bottom damage was also discovered after successive seasons and baffled us for a while, until it was realised that ice forced under the bow at speed would, after being pushed below the surface, rise and pummel the bottom plating in that area. Lloyd's Register of Shipping surveyors were not as happy about the extent of this underwater damage as our friends of the press had been about our dramatic but less serious damage to the topsides; indeed they were greatly concerned, and as we were insured at Lloyd's our premiums rocketed. Subsequently we annually met all the requirements to retain our Lloyd's Ice Class 2–3, for work in Light Baltic Ice, but carried our own insurance, backed by the government. However, as we gained experience of our likely areas of damage, we well exceeded the Lloyd's requirements in those places.

We three deck officers, each accompanied by an engineer officer, divided the period of refit to stand by the vessel and liaise with the shipyard into just over one month apiece. Adam and I bought a sturdy 1896-built gaff-rigged cutter of 36 feet in length, aboard which, berthed on the River Hamble, we lived whilst we each took our turn of duty. At the weekends during those periods we sailed with friends, but during the time we were both free we would undertake a more ambitious voyage just together. An amusing incident occurred on one such trip when, after getting caught in a nasty blow off Portland we put into Brixham to dry out, and the harbour master, whom we knew, managed to squeeze us onto a berth. The local regatta was shortly to commence, and a cub reporter interviewed us. On the basis of seeing our *Shackleton* issue jerseys draped over the boom, he reported that she was visiting to take part in the regatta. He had not seen the name of our yacht, *Marishka*, it being covered by soaked garments.

Also on our time off, encouraged by the Hydrographer of the Navy, Admiral Sir Edmund Irving, we both studied and completed a hydrographic survey course with the Royal Navy at Chatham. This course turned us into what the navy termed a 'charge surveyor'. It brought our surveying capabilities to a standard whereby the hydrographer could publish any survey that we were in charge of, as part of an Admiralty chart, or even as a complete one. This vastly improved standard of work was, of course, beneficial to ourselves, but importantly in the overall context such work could, when published, further display British sovereignty. We consequently equipped the ship with many more and better surveying instruments, to carry out such tasks. We also fitted our motorboat with an echo sounder which would produce a continuous trace of the bottom soundings. Prior to this we had possessed nothing more than hand leads to obtain individual soundings, which may have led to peaks or troughs in the seabed being missed, plus our sextants, a measuring tape, and a graduated pole. 'Eg' Irving, as he was affectionately known by all, a charming, witty man of great character, on retirement from his post as Hydrographer of the Navy represented the Survey on the Natural Environment Research Council, which we later became part of, and had much involvement and influence upon our affairs. Over the years we became good friends and I had the pleasure of taking him south.

There were during the summer one or two changes in our ranks. Unfortunately our electrician, although a humorous and likeable colleague and usually good at his job, drank too much too often, and had to leave. His replacement was purloined from the *Shemara*, Sir Bernard and Lady Docker's yacht that also used to refit at Thornycroft's Northam yard. His tales of events aboard the *Shemara* were prodigious, but whilst in the yard we witnessed her ludicrous departure as she was ordered to leave, stern first from a corner berth, where shipwrights on scaffolding were putting the finishing touches to the overhanging bowsprit. To their astonishment it slid cleanly from within their grasp. Our own departure, with repaired and painted hull, now showing no signs of those numerous indentations, was rather better planned by our new master, Captain Norman Brown. We went first for sea trials, thence to load, before sailing south on 1 October 1956. Norman Brown had sailed with Captain Johnston for some seasons, and eventually as his chief officer, but now took command of the *Shackleton* whilst Johnston went to Paisley to commission the new *John Biscoe* from the Fleming & Ferguson yard on the Clyde. This cycle of a long southern voyage followed by a short northern summer passing seemingly quickly, being packed with refit duties, periods of leave and initially studying for certificates and on courses, was to be the pattern of my life for almost the next 20 years.

4 RRS *Shackleton*: Second Voyage

Relief of bases and royal visit

The first part of my second voyage south aboard *Shackleton* took us initially, as was to be the case on nearly all voyages, to Montevideo and Port Stanley; then, on this occasion, onward for my first visit to South Georgia followed by a tour of the northern bases. Montevideo was always to be a pleasant stopover, synonymous with a warm sun and friendly people, good food, particularly charcoal-grilled beef, fresh fruit and salad, and cheap wine, all particularly welcome when returning from the Antarctic. But on this, our third visit to the port, our view of the laid-back, apparently innocent nature of this capital city began to alter as we recognised the underlying corruption and intrigue that was rampant there. The burnt-out shell of the dockside post office where we had deposited our letters twice on the previous voyage bore witness to the story that for a year or more no mail entering that post office had ever left it again, the employees, similar to those in many other of the Uruguayan Government departments, having done nothing other than collect their wages. Ultimately they had set fire to the entire building, including its contents, to clear the backlog.

During our stay the bronze propellers of a small Uruguayan naval vessel, berthed on an adjacent quay, were stolen from underwater! Suspected of being complicit, we were interrogated by the police. More tales of Montevideo arise from later visits, but finally on this call, our departure was delayed because several Fids had been jailed overnight for rowdiness in a dockside bar. Turned out at dawn to sweep the streets, they were not returned to the ship until their task had been completed. The young crew, led by their more experienced members, knew which bars to steer clear of around midnight to avoid this well-known entrapment by police patrols. We shuddered to think what opinions our owner might have hearing that her Royal Research Ship had nearly been impounded for being involved in stealing the propellers from the warship of a foreign navy, and half her scientists put in jail for drunkenness. Years later I learnt a trick from Captain Johnston that could be used, though none too often, when faced with false accusations or hypotheses leading to implausible and impossible demands in order to extract bribes, usually in the form of whisky. It was to mumble comments indicating that the owner was a lady of the very highest standing. As an awareness dawned amongst the handful of bandits, alias police, customs and immigration officials, the second stage of the farce would commence. The ship's register would be produced, which displayed our ownership in the name of Her Britannic Majesty, thumb carefully concealing 'represented by the Government of the Falkland Islands'. The ruse had a magical effect, for Uruguayans are not only very pro-British but also hold Her Majesty in great esteem. Finally escaping the avaricious clutches of the Montevideo officials and the amorous, though still pecuniary ones, of the strikingly beautiful girls there, we left Montevideo and the muddy waters of the Rio de la Plata, and headed for the Falklands.

Stanley, South Georgia and the South Orkney Islands

On this, our first visit to Stanley this season, we stayed for 14 days, unusually long compared to later calls. We had cargo to transfer to our stores depot and sister ship, and there were building materials for the erection of a base and a prefabricated refuge hut, to take aboard. We also had to await the arrival of further stores for the bases shipped directly from London on a vessel chartered by the Falkland Islands Company. Once ready to sail south with all but our deck cargo, we initially made the 50-mile passage to Rincon Grande, the settlement at the head of Salvador Water, a major inlet on the north of the East Falkland island, to load mutton carcasses. They were destined for the government station on South Georgia that we were to visit prior to heading to the bases. Returning to Stanley, we then loaded drums of kerosene, aviation spirit, acetylene and hydrogen bottles, and other dangerous goods on deck, topped off with timber, sledges and other lighter equipment that could withstand the weather, and then sailed, looking more like Steptoe's cart than a Royal Research Ship.

A quiet passage was made to Grytviken, where Argentina de Pesca operated a busy whaling station on the shore of the bay beneath Mount Hodges. On a small hillock to starboard on entering stood the very prominent white commemorative cross to Sir Ernest Shackleton. His grave was in the cemetery amongst those of whalers and seamen, beside the whaling station, for it was here at Grytviken, aboard the *Quest* in 1922 that he had suffered a fatal heart attack. Also to starboard was a spit of land, King Edward Point, upon which there were several bungalows housing the government officers, and a small wooden jetty at which we berthed. The magistrate/administrator headed a small team that included a customs/shipping officer, for during the season there were some 3,000 personnel based on the island, operating at this and the other three whaling stations. There was a meteorologist and a policeman, and indeed a jail. Although the stations were alcohol free, there was a great deal of illicit distilling of spirit from the likes of boot polish. Liquor was also acquired from visiting ships, often in exchange for engraved whales' teeth, scrimshaw work, causing drunken disturbances at times. Our ships' captains were also magistrates of South Georgia and the Dependencies, and could, amongst other penalties, lock people up for disorderly behaviour. The only incumbent of the jail I ever had to incarcerate was Duncan Carse, the actor/explorer who played Dick Barton on the radio, and who once made a one-man expedition to the island. He enjoyed a tipple when idle, and would make himself a nuisance aboard a visiting ship. Consequently he had to be shut away for a few hours to cool off.

There was no FIDS base here at the time, so our stay was only for a few hours to make our deliveries. For that reason and for fear of an avalanche, we did not on this occasion walk the beach path beneath the precipitous scree and snow slopes around the bay to see the whaling station. Our awareness of avalanches was very much heightened by the recent death of Ken

Butler, the whaling station manager, from such an incident when returning from a visit to the administrator one evening. He had been a Fid with Fuchs at Stonington Island in 1948.

Soon we were under way again towards the bases, Signy Island within the South Orkney group the first to be visited. Rounding the eastern end of South Georgia in a strong gale, with the sea pounding on Clerke Rocks, actually substantial islands which had among their group some islets humorously named Nobby and the Office Boys, here, as was often the case, we met a multitude of bergs, some still afloat and under way but mostly grounded. Their origin was the ice cliff shoreline of the Weddell Sea, some 800–1,200 miles to the southward, which they had calved from, possibly as enormous tabular bergs. Drifting in the clockwise current of that sea, they may well have travelled twice that distance before grounding in the shallows of this island's extremities, eventually to break into a myriad of more minor but still very large icebergs. As Cape Disappointment faded astern – named by Captain Cook in 1775, when he recognised that he had only discovered an island, and not the fabled *Terra Australis* that he sought – we came upon my largest berg to date. Having weaved a tortuous path amongst the earlier fragmented bergs and their associated bits and growlers, the latter often very difficult to see in the seaway, we could see this monster visible almost from horizon to horizon. By nature of its irregular shape, and observing only the face presented to us and upon the radar, we could only guess at the exact size of it, but certainly in excess of 25 miles across one axis. It was some 200 feet above the water and about five times that below, and extraordinarily uniformly flat-topped; we conjectured upon its total weight and how long it would take to disintegrate. As twilight fell and the wind dropped, Mount Paget, just short of 10,000 feet high and 100 miles distant on the quarter, caught the last rays of the sun, its summit tinting a rose-pink. The sea and sky merged into a single opaque entity enveloping and merging with our Goliath, which as if by magic disappeared from our sight, bringing the end to this spectacular day.

Neither South Georgia nor the South Orkney Islands had been on our previous season's itinerary, so I was sailing in the Scotia Sea for the first time. It seemed a grey, stormy place with much loose ice, and rarely any sun to lighten it, lying in the latitudes of the circumpolar trough[17] and thus directly in the path of the majority of depressions. Yet, because the coasts of South Georgia and the South Orkneys are the breeding grounds for so many birds, their proliferation across this gloomy stretch of water was a welcome relief when passage-making there. Similar were the whales, frequently noticed when blowing, whom we would dearly have liked to warn away from these killing fields. Many people aboard had read the accounts of Shackleton's boat journey during this passage, marvelling at their achievement in the 24-foot *James Caird* across these waters.

We made our landfall on Powell Island in the South Orkney group, one of the two islands of about 1,000 feet in height lying adjacent to the Washington and Lewthwaite Straits that in turn separate the two major islands of the group, Laurie and Coronation. The straits were another trap for bergs and their debris, which made for some incredibly tortuous navigation. The main islands have an east/west axis of about 70 miles and rise to about 3,000 and 4,000 feet respectively. Heavily glaciated, they make a fine sight when approached in good weather.

---

17 A band of atmospheric low pressure around the continent.

Discovered by the sealer George Powell in 1821, accompanied by the American Nathanial Palmer, they were named by James Weddell who, when ice conditions were incredibly favourable in 1823, struck so adventurously far south of them, to allow him to penetrate that sea which now bears his name.

The next important visitor deserving mention was Dr W. S. Bruce, who established a scientific and survey station on Laurie Island in 1903. I believe him to be a rather neglected man in the history of Antarctic exploration, although it is said that whilst he was a good leader, dedicated to his purpose and single-minded, he was a difficult person with whom to deal. Neither was he the best self-publicist, which may account for the little that is complimentary that has been written about him. He cut his teeth whaling, visiting the Falkland Islands in 1892–93, then developed as a naturalist and was offered such a position by Scott in 1900. This he declined, frowning upon Scott's main objective of reaching the Pole rather than his expedition being a scientific one; but also wishing to lead his own expedition. This he did, but being rejected for financial assistance by the Westminster government, he determined to finance his expedition in Scotland and made it the Scottish National Antarctic Expedition. His aims were to penetrate and survey the Weddell Sea, chart some of its coastline and survey and study the wildlife of the South Orkneys. He purchased and renamed as the *Scotia* a very strongly built Norwegian whaler, a three-masted auxiliary barque which was found to be in very poor shape. It cost him dearly to refit and modify her, but the resultant vessel was one of the best, if not the best, polar research ships of her time.

He created a base on a bay on the southern shores of Laurie Island, where he carried out meteorology, magnetic investigations and biology. During the winter of 1903 the *Scotia* overwintered at Laurie Island, and then in the following summer penetrated to 74 degrees south in the Weddell Sea, determining seabed contours. Dr Bruce also established the existence of the shoreline, linking it correctly with Enderby Land which lay to the eastward, calling his discovery Coats Land after the Coats Brothers, thread makers of Paisley, his original and one of his major sponsors. To achieve this he overcame considerable difficulties with ice and weather, and became entrapped for a winter by the pack ice at his Laurie Island base, which was nearly washed away by an onshore storm during his voyage to the southward. He had three men in his party from Argentina which indicates a liaison with that country, and because of lack of sufficient support from the UK government, on leaving his base in 1904 he handed it over to them. Their continuing occupation from that date until the present, keeping meteorological records throughout and other intermittent scientific observations, makes this the longest continually manned station in Antarctica, and led Argentina to claim sovereignty of the island group by way of that occupation. Scientifically, the keeping of those records and the baseline it sets is of tremendous value in the context of climatology. Dr Bruce went on to establish the Scottish Oceanographical Laboratory. He has nothing major named after him other than two small islands, but the mournful sea we had just crossed to the northward of this island group bears the name of his ship, the *Scotia*, as does the island arc which envelops it and the bay at Laurie Island in which she wintered. Outrageously – because, I believe, he worked outside the orbit of the Royal Geographical Society and the sponsors and supporters of Scott, particularly Sir Clements Markham – despite his expedition being an outstanding success, particularly scientifically, he was never awarded the Polar Medal.

We landed a small summer party on Fredriksen Island to study seal and marine biology, and whilst doing so came across a handful of fur seal, a rare find, for they were almost extinct. We then made our way to Signy Island, which lay immediately to the south of Coronation Island. Within Borge Bay, on its east coast, was a tight but comfortable inner anchorage, with close and protected access to a small wooden jetty, originally built by whalers around a sunken barge. This inner anchorage, Factory Cove, had been named and used extensively by whalers between 1910 and 1930, and remnants from their activities could be found abundantly along the shoreline. The station we had now brought stores to was the successor to a depot hut placed on the islands in 1946, it becoming a permanently occupied site the following year. Dr Richard Laws, a young scientist amongst the initial incumbents, initiated an important and continuing biological programme. After that initial southern foray, he spent time in Africa before he followed Sir Vivian Fuchs as our director in 1973, holding the post for my last year with the Survey, and then until 1987.

Signy Island is an altogether more visitor-friendly island than its larger neighbours. Hardly glaciated except for a small permanent plateau ice cap, its highest point just short of 1,000 feet, it has only minor peaks of little significance. Apart from steep cliff faces it can therefore be traversed relatively easily, and we were well rewarded walking and scrambling over it, perhaps with too much enthusiasm, as we learnt over the ensuing years. The island is an outstanding example of the Antarctic maritime region. In summer a great deal of bare rock emerges from beneath the snow. As the snow melts from the rock and the snow slopes, it turns beautiful shades of pink from the snow algae within it. Much of the bare rock then reveals lichens and mosses. My knowledge of the flora has been gleaned primarily from listening to Dick Laws, so I believe I can do no better here than to quote from his splendid book *Antarctica – the Last Frontier*. He writes:

Half of the ice-free surface of Signy Island is covered with lichens, and mosses are quite common, but Antarctica's two flowering plants are less abundant. The low growing lichens on the sea coasts, some of which can live 4000 years or more here often grow in colourful stripes, from black and brown around the bases of the cliffs to bright yellow and orange at their tops. Among the mosses here, the tall moss turf is one of the most striking forms of Antarctic vegetation. This forms peat banks up to three metres deep, whose bases regularly date back more than 5000 years. Even apparently dead black moss peat, which has been buried by ice for a century will regenerate when it becomes exposed. Because all these plants forms grow very slowly in Antarctica's cold conditions, they are very vulnerable to disturbance, particularly trampling; on Signy Island there are human footprints clearly visible decades after they were made.

In hindsight, fortunately, a great deal of our exploring the island on this and subsequent occasions was done by initially taking a ship's boat to find a quiet cove, land and thence trample ashore. Nevertheless, I dread to think how many footprints we left for posterity. Freshwater lakes enriched by the detritus from seals and birds added to the interest of the flora and fauna. There were rookeries of chinstrap, macaroni, Adélie and gentoo penguins. Large and small petrels, cape pigeons, sheathbills, skuas, shags and terns were all found nesting, and taking a trip at dusk across the Normanna Strait, the mile-wide separation between Signy and Coronation Island, we witnessed the incredible mass exodus of prions from their nesting holes on Cape Hansen, to feed at sea. Of the many seals, always intriguing and entertaining were the gargantuan elephant seals, wallowing and belching, the bulls inflating their proboscides and fighting, usually a young male invading an older bull's harem of cows. Within the base the marine biologists had set up holding tanks full of a variety of creatures and insects collected from beneath the underside of pack ice and from the seabed; it was a revelation to learn what lay beneath the surface of our often apparently barren sea. The weather held up well during this, my first visit here, and allowed our work to be done without interruption, leaving us time to enjoy the magic of a very special part of the planet. There was no hint of the difficulties in appalling conditions which I was to be faced with some years later.

This time we wished we could have lingered longer, but an impending visit by Prince Philip aboard the Royal Yacht *Britannia* to Port Stanley on 5 January had imposed an unusually strict timetable on us at this, the start of the season. This was to be a whistle-stop tour of the northern bases before we played our part there, which did not please many aboard. We had become dedicated to our work and enthusiastic about our environs south, and believed the coming involvement an annoying distraction to our limited season. Ultimately we thoroughly enjoyed our participation in what was a grand occasion for the Falklands.

The passage towards our next base, Admiralty Bay, was interrupted by the ship's cook falling ill. We turned to retrace our course back to Signy Island, where they had a resident doctor on base to care for him, for at this time we were without one. But because the condition of the cook deteriorated severely and the base doctor was unsure of his ability to cope with the equipment he had on base, we again altered course, this time for South Georgia, where the Salvesen whaling station had a medical facility where he could be operated upon.

We made the 800-mile crossing of the Scotia Sea once again to Leith, where we berthed for little more than an hour, landed our sick cook and then departed for the South Shetland Islands, this time going west about the island and cutting south through the Stewart Strait. With its, dark, ominous landscapes of surrounding islands, devoid of any snow at this time of year but shrouded in mist and rain as they caught the brunt of the westerly weather, and in a fiercely heaped sea, as a strong tide with us ran against a south-westerly gale, it was a fearsome place. Yet the whole seascape, nature in its rawest, was full of weaving, diving albatross, giant petrel, skuas and smaller birds from their nesting grounds on the nearby shores, particularly Bird Island, named appropriately by Captain Cook. As we moved into the open sea the wind, having been funnelled through the strait between the lofty islands, lost some of its force but remained a strong westerly from a couple of points on the bow, making for an uncomfortable passage towards King George Island in the South Shetlands. Had we been able to continue our original voyage, and left Signy Island for Admiralty Bay without deviation, we would almost certainly have got south of the westerly winds, and possibly even had a favourable easterly to make an easier passage across the top of the Weddell Sea, albeit with the chance of meeting more ice. As we passed majestic Clarence and Elephant Islands, the weather cleared; we obtained both good sun and star sights and found the islands some miles out of position on our charts. Also in the vicinity we found a previously uncharted seamount of interest and of some navigational consequence, for we utilised such obvious variations in the seabed to help position ourselves when other means were not available.

Little time was spent at Admiralty Bay, however, as we began a pattern of frequently sailing in late evening, after we had finished cargo work and had provided some succour and entertainment for the base Fids, which made for a hectic schedule. With the sun above the horizon for almost 24 hours, working all day, keeping a four-hour watch during the night and catching some sleep either side of it meant we were busy, but it was no hardship. On that occasion, with the beauty of the mainland looming ahead on which to make a landfall and the variety of ice to be negotiated, and wildlife to be seen, such passage-making was a pleasure, and the 100-mile transit of the Bransfield Strait seemed to pass quickly, bringing us to Hope Bay.

Hope Bay

The South American Andean chain enters the sea at Cape Horn and travels firstly east forming the Burdwood Bank to the south of the Falklands, then onward still in that direction to appear above water as Shag Rocks and South Georgia, thence south-eastwards re-emerging above water to form the South Sandwich Islands and curving to the west again, re-appearing as the South Orkney and South Shetland Islands. The whole island and subterranean feature is known as the Scotia Arc, and contains the Scotia Sea. From the South Shetlands, the arc plunges southwards to re-appear as a continuous mountain chain forming the spine of the Antarctic Peninsula, and then onward into central Antarctica. At the tip of that peninsula, the most northerly point of Antarctica, and at the very foot of that final magnificent up-thrust of mountains, lies Hope Bay.

Dominated by Mount Taylor, a snow-capped peak of 3,200 feet, Hope Bay, a two-mile indentation of the coast, and its immediate surrounds, displays its association with the Andes, and provides its introduction to Antarctica, with flamboyant pride. For here, very many of the stunning features that epitomise the continent can be seen. A further two mountains, the 1,400-foot Whitten Peak and the 1,700-foot Mount Flora, rise from a pristine ice slope, itself attaining about 1,000 feet and flanking the snout of the dramatically crevassed Depot Glacier. Two considerable lengths of ice wall, varying in height between 50 and 100 feet, border much of the shore, whilst the remaining rocky and cliff coastline contains a central stone beach, which provides a good landing place, before, within a few yards inland, the ground becomes ice-clad. In the spring the shoreline here can be an ice foot; winter sea ice joins the land-borne snow and ice to create a level platform across the shoreline, but whereas the land ice stays firm, the sea ice rises and falls on the tide. This creates a fissure, a tide crack, and when the sea ice moves out in the spring, it leaves behind a short face at the tide crack, an ice foot. Whilst this remains intact, it provides an excellent quay-like landing on the right tide. Unfortunately here in these latitudes, as summer progresses, it usually erodes and the sloping beach itself had to be worked. The blissful scene was completed by an Adélie penguin colony of several thousand pairs. The species was named, with his tongue in his cheek, by the 19th-century French explorer and navigator of this region, Dumont d'Urville, after his wife Adèle, it being the most belligerent, aggressive and chattering of the penguin family.

Regrettably this extravaganza of natural beauty was not always to be so magical! The Fids' base hut, called Trinity House, had a mixed history, and the bay, whilst so attractive, a fearsome reputation. Post-war after its Operation Tabarin inception, Hope Bay became by 1947 primarily a sledging base to further topographical survey and geology down the peninsula. In that year, after some depot laying of over 350 miles, Ray Adie, later to become deputy director, and Frank Elliott, the Secretary of Fids in Stanley at this time of my second voyage, made an outstanding journey from this base of over 600 miles, overcoming almost every imaginable polar obstacle to reach Stonington Island base, then the most southerly British station. Sadly the elation at such success was short-lived when the following year the base hut was burnt to the ground with the loss of two men.

The base was reopened by the *John Biscoe* in February 1951, only for their cargo landing party to be shot at by the military personnel of the Argentine station that had been sited there during the short British absence. A Gilbert and Sullivan episode followed, with the governor arriving offshore in HMS *Burghead Bay*, Royal Marines storming ashore and the Argentineans fleeing to the hills. All eventually to be sorted out by quiet diplomacy between London and Buenos Aires, and the governor and the Argentinean base commander both supposedly being reprimanded for their hastiness. A further fire, fortunately without loss of life this time, took place in 1960 destroying much of the base. Afloat, weather and ice played havoc with any British intentions right from the start. In 1944 the *William Scoresby* and *Fitzroy*, the latter on charter from the Falkland Islands Company, had attempted to establish a base here under the leadership of 'Boy Scout' Marr, who had previously gone south with Shackleton when only a 17-year-old. Conditions were so atrocious that the ships left the area with only half the requisite materials and stores landed, after the *Scoresby* had attempted to anchor and work the new base site from 12 miles off. The base was erected the following season. The holding ground was rock, extremely uneven, with deep gullies and sharp ridges, and it was a matter of luck whether or not the anchor held. Katabatic winds down the glacier and the funnelling effect between the mountain peaks could turn the quiet bay into a seething inferno in moments. With bergy bits and lumps of shelf ice both afloat and aground, and pack ice moving in and out, leaving brash ice against the shore, it could be a very difficult place to lie and work. A few years later we were to be blown ashore here in 120 knots of wind, but this time, having successfully relieved the base, we left unscathed.

Deception Island

Re-crossing the Bransfield Strait, we called at Deception Island with its highly volatile weather, alternating between sunny calms and north-easterly gales of some days' duration. Hunting

Aero Surveys were already established for their second season, but as yet without their aircraft. The two Canso flying boats having reached the Falklands from Toronto, Canada were now awaiting a break in the poor weather both there and at Deception before crossing the Drake Passage. It had been intended that we would be at sea in that area when they were en route, to provide assistance in the event of their having to put down, but the earlier disruptions to both our own schedule, and that of their support team in the *Oluf Sven*, delaying the preparation of the Deception landing facilities, had spoilt that plan. The aircraft had actually arrived in Stanley well ahead of their team aboard the ship, and indeed rather too early to go south for the commencement of the Antarctic season. Consequently they had paused for five weeks, completing an aerial survey by vertical photography of the whole island group on a scale of two and a half inches to the mile. It was interesting to note, that of the five weeks during October and November, which is considered one of the better periods for weather in the Falklands, only eight days of actual photography was possible because of strong variable winds and cloud cover.

At Deception our offloading was disrupted by yet another severe gale, but worse nearly befell the Huntings team. They had a contraption, some folding boat equipment loaned to them by the Royal Engineers, nicknamed the Circus. It consisted of two pontoons, with steel transoms and trusses, and articulated ramps either end to enable heavy vehicles and equipment to be borne ashore. Carried folded in the hold of the *Oluf Sven* it had been put together and anchored a little way offshore. The gale sank it. On receiving Huntings' appeal to our ship for help, Adam and I were detailed off to raise it. We grappled for a few hours to find it then passed wires through various parts of its invisible structure. We then placed both motorboat and scow over the wreck, hauled tight and made fast the wires at low tide, and let the rising tide lift it off the seabed. We then moved our makeshift salvage rig with the Circus slung below towards the *Shackleton*, but within minutes everything parted and it sank into even deeper water. Fortunately we had some lazy lines attached, and before long we were able to pass heavier wires through its structure, and this time eventually under it, and then attach them to the ship's crane, she having shifted to nearly directly over it. We then gently eased the Circus to the surface, drained it and about midnight of the same day put the now tangled mess of wood and steel on deck, having commenced the operation at 0630 hours. The next day it was taken apart, transported ashore and re-assembled. As it was their only means of transporting heavy engines, generators, their three-ton tractor and other major items from ship to shore, they would have been sorely tested without it.

The gales eventually blew themselves out as the fronts of the disturbance moved eastwards from both Stanley and Deception, and Huntings called in the aircraft. Their passage south was not uneventful. During the 600-mile flight, severe icing occurred in cloud, and a lump spun off one of the propellers of one aircraft, punching an eight-inch hole in the after cockpit, only narrowly missing the captain. By the evening of their flight both aircraft were swinging gently at their moorings. Laying these, in which we had played the major part, had been one of the most difficult preparatory tasks before the aircraft could be received. With some help from the Huntings team, discharging and moving our cargo with their large tractor up to our base in return for our assistance to them, we completed and sailed. Setting course for Danco Island station, which we had established the previous season, we made a short diversion towards Hope Bay, from which two Argentineans had been blown in their dinghy when the outboard failed. The Canso aircraft also took to the air to carry out a search, and the Argentineans had a ship in the area, so our services were dispensed with and we soon continued with our voyage southward. Yet again my views on the use of dinghies down south were reinforced. The two men were lost.

On our way down the Croker Passage towards the Gerlache Strait, we again 'shot up' the obvious geographical features, and ran a line of soundings as we made good a slightly different course to that when transiting the same stretch of water on a previous occasion. The Canso aircraft had made flights over this area when commencing their aerial survey and even at this early stage, before their material was worked up, were able to confirm some of our more questionable major discoveries. Trinity Island was found to extend some ten miles more to the eastward than thought, and whilst Hoseason Island lay six miles to the west of its charted position, Two Hummock Island, only 20 miles further south, lay five miles to the eastward of its drawn position; these were all major islands bordering our route. The charts of the area, with the exception of a few northern islands and primary anchorages, were of course only based on the sketch surveys and notes of earlier explorers and whalers. Their efforts had been undertaken spasmodically and in isolation, and little attempt had ever been made to link the various pieces of work. Huntings, seeking to establish ground control for their surveys by triangulation, would make the first ever to attempt to link the South Shetland Island group with the mainland and its off-lying islands. We also found that the mainland coast appeared at times to have been placed by as much as 15 miles to the westward of its true position. The latter error could well have been on account of the observers mistaking fast ice or a multitude of broken ice against the coast, for the actual coastline. On our passage, we managed to complete a running survey, and with reassurance from the Huntings team above us, we confidently filled in our Whatman board, producing a workable chart.

Return to Danco, Anvers and the Argentine Islands bases

A year on at Danco Island, we found that nature had corrected the disturbances we had created when building the base. Deep snow lay all around the hut hiding any scarring of the landscape by stores boxes, building materials and oil drums, whilst penguins had returned to within feet of the building. Outside it looked pristine and charming, and in its island setting, surrounded as it was by the fabulous more majestic close scenery of further islands and the mainland, it was another magical place to be. But aware of the earlier failure to find a nearby route to the mainland plateau, we proceeded into Wilhelmina Bay to find a site for a depot hut, from which to achieve that objective, hopefully to be reached from the base by boat in summer and over the sea ice in winter. Wilhelmina Bay was virgin territory, and the chart, such as it was, bore little resemblance to that which we found. That led to some exciting incursions into shoal areas, around 'new islands' and into fjords that were supposedly straits but found to be without exits. We eventually found a site accessible from seaward, with possibilities of a route inland, and established a camp there, at which we left a climbing party. On our return passage north we were to place a hut there if their explorations had proved successful, or lift them if not.

We then sailed across the Gerlache Strait to Port Lockroy, where we paid a visit of only a few hours before heading to the Anvers Island base. We enjoyed the scenic passage rounding Cape Lancaster, backed by Mounts Hindson and Williams, but the last few miles approaching Anvers Island again proved a nightmare. Small islets, rocks, shoals and reefs barred our way, despite the assistance given us by the cairn we had erected on our previous visit. The base lay adjacent to the pretty cove of Arthur Harbour, named after the governor of the Dependencies at that time. Attractive and businesslike as it was, we realised that when those disciplines of mapping and geology for which it had been established were completed it would probably be abandoned, especially as it was barely 30 miles from Port Lockroy.

The little harbour nevertheless became a favourite place to lie, with swinging room clear of dangers and fairly good holding though, as in Port Lockroy, in a northerly gale blown snow streaming from above the ice cliff was aggravating. But there remained the perilous approach to be negotiated through the extraordinarily shoal-strewn last few miles from the deep and wide Bismarck Strait that we therefore attempted to survey. When each day's unloading had finished we set about identifying the shoals down our approach course, but found that the most problematical ones were too far from any positioned marks for us to chart them, particularly with ice limiting our view from only standing height within the launch. Consequently from the ship on departure, we recorded the vertical sextant angles and bearings of our cairn previously built on Torgerson Island, when close alongside the dangers. When these positions were subsequently approached, deviations could be made to avoid them before returning to our original base course, and thus a very rudimentary but satisfactory sketch survey was prepared to enable a safe entry and exit of the harbour. We used the chartlet thus created throughout my time, then provided it to the Americans when they later took over the base, and even passed it on to others to use after my retirement.

The initial impediment to this method of using a single bearing course and positioning of shoals was the unreliability of our gyro compass. It did not perform well in high latitudes, particularly when fast turns were made. Situated in front of the helmsman in the wheelhouse, it was thrown by large variations in temperature, such as when the windward wheelhouse door was opened, and was also thrown by the more violent shudders of the vessel on impact with ice. These problems meant that we were often forced to rely entirely on the magnetic compass. This could be seen by the helmsman, via a periscope through to the binnacle on the monkey island,[18] but it was not visible to the captain or officer on watch. Without the gyrocompass there were no gyro repeaters working on the bridge wings, and the magnetic compass having none, taking bearings with an azimuth mirror[19] was precluded; an entirely unsatisfactory state of affairs for the fine handling required to avoid those shoals.

---

18 Open upper bridge.

19 An azimuth mirror is a device fitted over a compass binnacle or repeater with which to take bearings, seeing both the compass and the object viewed simultaneously in the same direction.

The second piece of equipment to underperform was our echo sounder. The cold, more dense water inhibited the transmission of the pulse, as did any ice under the hull. The electronic depth recording by a continuous trace, made by a stylus marking a damp roll of paper that would often partially or completely freeze, resulted in the loss of any indication of the depth. When clawing about to search for a passage through shoal water at minimum speed, lowering the anchor by one shackle gave warning of a rising bottom when it touched at 90 feet. This would have been an impossible practice when passage-making at speed, as it would have risked tearing the chain out of the locker, and losing both it and the anchor. It would appear that we stumbled from pillar to post; but in fact unless groping amongst rocks for an anchorage, or standing in to an unknown coast to make landings, our progress on passage was, whilst cautious, mostly quite steady and ordered, without a hint to those not on the bridge of the number of times we up there held our breath. In hindsight I now believe that we were in a period between the basic exploration methods of earlier eras and those of the more modern and sophisticated shortly to come. We had so many instruments like the radar, echo sounder and gyrocompass denied to our forebears, but they did not work as well as expected, or indeed at all, in these latitudes and waters. The instruments needed refining and adjusting for our conditions, something that no one had given any thought to. With those navigational aids aboard, which worked satisfactorily in other parts of the world, we expected better performance of our equipment and therefore of ourselves, and consequently pushed the boundaries, only to give our small navigational team a whole set of original problems.

Leaving Arthur Harbour we travelled east in the Bismarck Strait then south towards Cape Renard, which when passed close to port led into the glorious Lemaire Channel. What a feature to bear one's name, and it was de Gerlache who gave to one of the outstanding scenic sites of Antarctica that of a Belgian explorer of Africa! Our next destination was Base Faraday on the Argentine Islands, and we utilised exactly the same procedure approaching Cape Renard, that of maintaining a single course on a bearing towards this cape, as we had just evolved at Arthur Harbour. Once more there were visible dangers, and shoals suspected close to hand, but our route, keeping Cape Renard right ahead, which had become standard practice in our ships, took us safely through, albeit crossing some striking peaks and troughs of the seabed as indicated by the very intermittent echo sounder trace. One only had to look at the twin peaks of Cape Renard, close at hand, soaring as they did some 2,000 feet near-vertically out of the water, to realise that there was no reason why such precipitous peaks of rock should not rise similarly from great depths, their summits lying as dangers just beneath the surface of the sea. Extreme caution had to be taken whenever the depth shallowed to less than 100 fathoms.

Clearing the southern end of the Lemaire Channel, we met heavy pack ice trapped tightly in the Penola Strait between the mainland and the numerous offshore islands. More than once we were brought to a halt, taking an hour or so each time to extricate ourselves; and only four miles from our destination were forced to give up and retreat, the master not wanting to risk becoming beset; two years previously, as chief officer of the old *John Biscoe*, he had been stuck here for some weeks and could not risk that again with duties in the Falklands not that far off. Having established by radio that the climbing party left near Cape Reclus had reached the plateau, we returned there and built a refuge hut.[20] That completed, we sailed south to make one further attempt to reach the Argentine Islands. For the second time we found the pack ice too consolidated, the *Shackleton* proving herself not powerful enough to force her way through any major concentrations. We completed the first third of this season by revisiting Deception Island and Admiralty Bay before heading for the Falklands. There were always personnel and stores to be transferred, and whilst in the area the masters of the ships, as father figures, co-ordinated general inter-base matters and requests for assistance. Each evening at ten o'clock all the bases and both ships would come on the radio for the Goon Show. Here, amidst a lot of pleasant banter, many of the minor arrangements within our programme would be discussed.

At Deception we also had a task to perform for ourselves. On the passage northward from the Argentine Islands bad vibration was experienced from the propeller, and we presumed that it had been damaged whilst working the pack ice. Having anchored, with the vessel trimmed by the head, it could easily be seen that one of the four blades had at least five inches of its tip bent at right angles to itself. It was decided to change the blade for one of the spares we carried on deck. To achieve this we put the stern ashore on the ash beach at high water. Fortunately it was spring tides, with a good rise and fall, and we chose a sloping area that would not put too much upward strain on just the stern section and would also support the vessel well along its length forward. When the tide fell we had the spare propeller blade hung off over the stern in readiness for the engineers as they prepared to remove the damaged blade. But we soon found that our carefully selected part of the beach was an area of very hot springs issuing scalding steam, which of course we had been totally unaware of when viewing it at high water. Ironically, these conditions, combined with pieces of pack ice floating about in the bay, made it quite impossible to carry out the work there. With better awareness of what to avoid we immediately chose a satisfactory spot whilst we could see the beach at low water, and then on the next high tide shifted ship. The blade was then very easily unbolted and the new one put in place.

A royal visit facilitates a close look at the Falkland Islands

We arrived in Port Stanley on midsummer's day for the build-up to Christmas and the arrival of *Britannia* in the New Year. There were endless parties thrown by and for the same group of ex-pats and local pillars of society, centred on those thrown by the governor. Probably typical of many colonial outposts, these social occasions were devoid of any serious conversation, fuelled by gossip and drink, and far removed from anything concerning the real world. For youngsters on a visit for some R and R between quite hectic periods of duty, it was great fun if sensibly interspersed, as it was by us, with some healthy outdoor activities and any work necessary aboard. We young officers, by nature of our intermediate position, had the ability to span and make friends across all sections of the local society. We partied with the elite, were mothered and fed by the wives of the colonial and company officers, and joined the Colony

---

20 The 'Reclus' hut is now at the Stanley Museum as a polar exhibit.

Club, drinking and playing snooker with the hierarchy, yet we drank in the pubs where we met the local girls and made friends with the down-to-earth and hard-living islanders. We fished, and shot hare on the slopes of Mount Tumbledown, and Adam and I sailed the ship's dinghy, which I had acquired in Southampton. Slung in a pair of small davits we had aft, it was sturdy enough for the Falklands weather. However, with the gusty squalls in the harbour we often capsized, and the residents of Stanley thought us crazy, whilst we wondered why we had maintained such a severe timetable down south to spend so much of our time now in the Falklands at play. The answer lay in that up until and at that time, FIDS was under the control of the Colonial Office, and thus the governor, and our modus operandi was that of colonial civil servants. There was a transition developing that we were not then aware of at that time; we were to become far more controlled by our professional director, with science and its support in mind, than by the governor, a layman in matters of the Antarctic, though still responsible for the area. At that time, though, we played like government employees on their summer holidays – but it was just about the last time we spent so many days practically idle alongside at Stanley.

The high jinks started on our very first night there with a fancy dress ball at Government House, to which we wore our uniforms! A ship in port was always an event in Stanley, an excuse for a drinks party, and in return we hosted our own event for a favoured few, followed by a meal and a film show. On Christmas Eve, Adam and I abandoned partying and sailing for flying with the local air service, and made a tour overhead of many of the camp stations to drop, literally, their last yuletide mail. We spent a lively Christmas Day in the home of Captain Freddie White, who was master of the Falkland Islands Company ship RMS *Darwin*. Freddie, a fine seaman who was larger than life, possessed a great knowledge of the waters around the Falklands with the extremely strong tides running through the rock- and kelp-filled passages between the islands. Coming to the Falklands in an Eagle Oil tanker, he had met and married a local girl, Nell Pitaluga, a descendant of one of two brothers who had come to settle in the islands in the early 1800s, and whose relations then and still now farm those original holdings. Freddie was very good at networking, and far from being isolated and out of touch down there, he became very well connected. During his 20 years plying between the islands and mainland South America, he became friendly with Captain George Barnard of Blue Star Line. Barnard became an Elder Brother of Trinity House and eventually Deputy Master (Chairman of the Board). Freddie then succeeded in becoming an Elder Brother himself, and moved to the UK to join the board. Sadly, after only a few years in office, where he had been a breath of fresh air, and I myself had spent only six months in his company on that board, he died.

The *Britannia*, with the duke on board, arrived from New Zealand at the anchorage surveyed for her near Port Lockroy on the Antarctic Peninsula. There he transferred to the new *John Biscoe* on her maiden voyage. Captain Johnston then took HRH on a tour that included six bases. With the royal timetable in mind, but wishing to show the duke as much as possible in the time available, he cut through an area of shoals and islets to the westward of the peninsula where no ship had yet ventured. In the process, the *Biscoe* glanced off a rock, Johnston purporting to a ruefully smiling Prince Philip that it must have been a lump of ice that he had hit. Later, when we all met up for a debrief on the duke's visit, Johnston announced that the only thing that had gone wrong was that 'sodding rock' getting in his way. To this day it is known to Fids mariners, and appears on our surveys, as Sod Rock, though unfortunately it was never formally endorsed as such. Re-embarking aboard *Britannia*, the duke crossed the Drake Passage to arrive in Port Stanley on 5 January 1957. We in *Shackleton* were to be there for his arrival, but first had an interesting task to perform; we took off around the islands making short calls at innumerable settlements to embark those from the camp[21] wishing to be in the capital for the occasion. For the ship's company, particularly the deck officers, with the intricate navigation involved it was a marvellous opportunity to venture beyond Stanley.

The Falkland Islands had been discovered by John Davis, a mariner from Waddeton, on the Dart Estuary in Devon. Commanding the *Desire*, he left Plymouth with a small fleet of several ships to explore the Philippines and China via Cape Horn. Driven from the coast of South America by storms, they came upon the islands in August 1592, and took shelter. Then in 1690, a Captain Strong made the first recorded landing, and named the straits between the main islands after Lord Falkland, the Treasurer to the Navy; by 1745 the name had been attributed to the whole group. The French name for the islands, 'Les Isles Malouines', has its origin in St Malo, from where de Bougainville sailed in 1764 on a voyage during which he established a colony, St Louis (now Port Louis) in Berkeley Sound, and claimed the islands for France. The Spanish interpretation of that name gives rise to the Argentinean name for the islands, Islas Malvinas.

At about the same time as de Bougainville's arrival, an Admiralty expedition created a successful settlement at Port Egmont, in the north-west of the islands. Whilst there, they came upon the French at St Louis and told them to leave. This upset the Spanish, who felt that they were in control of matters South American. Britain, Spain and France remained at loggerheads over the issue until France received a payment from Spain and withdrew her colonists. Spaniards resettling St Louis renamed it Soledad, whilst the British settlement at Port Egmont grew, until the Spanish sacked it in 1770. Britain and Spain nearly went to war over this episode, but the dispute was settled in 1771 when Britain re-established its ownership of the islands, and resettled Port Egmont, only to abandon it in 1774, leaving the colours flying and a plaque declaring British sovereignty. The Spanish remained at Soledad until the early 1800s, when they too withdrew.

From then the islands appear to have been unoccupied until 1824 when a Franco/Argentinean adventurer settled at Soledad. He foolishly captured three visiting vessels of the United States, which brought the wrath of America down on him in the form of the USS *Lexington,* which destroyed his fort and retook the US ships but left the settlement standing. In 1832 HMS *Clio* arrived, expelled those Argentineans remaining and, once again hoisting our sovereign's flag, established the British Crown Colony. The story, however, does not end quite there. The inhabitants of Port Louis were a rough mixture and they murdered the British governor and his assistant. A new administration had to be formed there in 1833, which was

---

21 From the Spanish *campo*, the 'camp' is the countryside in the Falklands – that is, anywhere except Stanley and the air base.

later moved to Stanley, no doubt because it was an easier harbour to sail into, and there it remains today, whilst the issue of ownership of the islands is still raised by the Argentineans today despite their trouncing and removal after the 1982 invasion. Has the Argentine any rightful claim to the islands? Significant is that in 1833 the Argentine state did not actually exist.

The Falkland Islands comprise two main islands and some 200 much smaller ones, all remarkably similar to those off northern and western Scotland. Both main islands rise to approximately 2,300 feet, with one significant mountain each and with many lesser, though sharp and rugged, peaks and outcrops of rock. Treeless, barren moor best describes the terrain, peat bogs being frequent. No rivers of consequence exist, but the coast is greatly indented, long arms, necks and inlets running inland from the sea, and many bays and sounds affording good harbours and compensating shelter from the often fierce weather. There are many interesting passages in which there are very strong tidal streams, between the innumerable small islands. Kelp, (Sargasso weed) grows abundantly in depths of less than 30 fathoms and assists navigation by marking shallow ground, but can be run under by the strong tides, giving the false notion of safe depth when invisible.

The weather, as is to be expected with the island lying in the South Atlantic some 300 miles east of the Straits of Magellan, is essentially maritime. Both depressions and anticyclones move quickly on fast-flowing upper air, bringing gales about once every five days in both summer and winter, not particularly cold except when an occasional southerly feeds cold air up from the glaciated shores of the Antarctic, which lies less than 700 miles to the south. Not much fog occurs, either, but close offshore poor visibility is frequent, caused by low cloud and rain against the coastline. Perhaps unkindly but truthfully, wind and rain sum up the islands' weather, yet they are a pleasant and interesting place to visit. The sun is strong when it shines on a still summer's day, or if one can find shelter, for there is no pollution and the islands lie at only 51.5 degrees south, equivalent to the latitude of London in the north.

Despite their apparent repose on any atlas, the islands do not form part of the Scotia Arc, the extension of the Andean chain, nor are the islands similar geologically to Patagonia, against whose continental shelf they lie. The Burdwood Bank, lying only 90 miles to the south of the Islands, does form part of that arc. When Gondwanaland, the vast southern supercontinent of which Antarctica was the keystone, gradually broke up between 180 and 25 million years ago to form the separate land masses of Antarctica, Australia, South America and Southern Africa, the Falkland Isles spun off from adjacent to what is now South Africa. The rocks of the West Falklands are Devonian, and those of the East a mixture of that and Permian. For the layman, though, the most striking geological features of the islands are the stone runs or rivers of stone, for that is exactly what they look like, to be found spilling down many of the valleys. A host of theories exist regarding their formation; Darwin thought the origin of these quartz-like boulders to be volcanic action; whatever, they are very unusual.

As is so often the case with wild isolated parts of the planet, it is the wildlife of the Falklands that is one of its main attractions. It is varied and abundant. Several types of penguin breed around the shores, from the large king penguin to the smallest of the breed, the rockhopper.

Some, particularly the jackass, unlike their Antarctic compatriots whose rookeries are on exposed rock, shingle or ice, nest in burrows amidst the boggy foreshore. These humorous little creatures, in their muddy, soil and sand environment, are sometimes somewhat sad in their dirty appearance, compared with their pristine colleagues further south – although they too can get messy and smelly, waddling in their own guano on a warm day. The black-browed albatross, one of this enormous but graceful species, with its long lumbering take-off belying its oceanic passage-making ability, leads the array of birds to be seen. There are turkey buzzards, carancho, kelp and upland geese, and shags. The latter is the sailor's friend, for if you are making the land in poor weather and sight a shag, you can be assured you are within seven miles of the shore. On the other hand, many of the birds – black-necked swans, oyster catchers and gulls are but some of them – are disliked by farmers for eating the sparse grass, as are the birds of prey for their habit of attacking fallen sheep or weak lambs.. Of the ducks, the logger or steamer duck is to be found in all inlets and makes for happy viewing as they chug along with their offspring amongst the kelp. Unfortunately, their eggs, together with those of the penguins with their lurid orange yolks, made a welcome addition to the Falkland Islanders' breakfast of mutton chops. Upland geese, too, make very fine eating, as do their eggs. Cattle farming was the original occupation of the settlers; de Bougainville brought them from South America along with other animals, but sheep farming took over later and was in my years the main stay of the economy; neither offshore fishing nor oil nor tourism featured as a possibility then. Sea lions, leopard and elephant seals are plentiful, and fur seals were regaining their numbers, recovering from their slaughter in the 19th century. Magellan porpoises are frequently sighted around the coast, and often accompany craft entering harbour, riding their bow wave or playing alongside. Tussock (Tussac) grass, up to six foot in height, is prevalent on the coast of the main islands, completely covering some of the smaller. It collapses annually, but in doing so and regrowing forms the tussock bogs so beloved by seals and penguins.

As we progressed around these fascinating islands we became enthralled by their surprisingly enchanting appeal, the intricate waterways separating them and the wave crests of a good breeze in the open reaches sparkling in the intense light, whilst towards the shore masses of long, brown, glistening fronds of giant kelp encircled pools of mirror-still water reflected the sky, undisturbed except for the foraging of a logger duck. The black tideline of weed and crustaceans fronting the green tussocks, brown bogs, yellow grass and grey outcrops of rock ashore, with the many bright autumn hues of a good October day in Britain from the diddle-dee and fachine bushes, and the gorse, wild strawberries and flowers beyond, produced a kaleidoscope of colours as far as one could see. Then quite suddenly, the whole scene would become completely obliterated by a passing squall, all to re-appear once more; and almost certainly on many days that process was frequently repeated.

The settlements off which we anchored, running our motorboat into the little jetties, mainly consisted of the Big House where the manager or owner lived, some smaller houses for married couples, and the cookhouse where single men lived. The latter was also the social centre of the settlement, whereas the wool or shearing shed was the working hub. At large settlements there were both a small store for purchases and a school. We embarked 184 passengers, most of whom had never been into Stanley before, having lived their entire lives in the camp. Most of them were shouldering carcasses of lamb or mutton, and just about all had a bottle of rum in each side pocket of their jacket as presents for the relatives that most of them had in town, and with whom they would stay. On board, they slept in any spare cabin, throwing down their bedding and saddles in the mess rooms and alleyways, but more often than not just enjoyed one massive party.

As we returned to Port Stanley some saw for the first time the white quartzite stones on the northern hillside spelling out BEAGLE, which had been laid out by her crew in 1834; Fitzroy was her captain, with Charles Darwin aboard. The whalebone arch erected by whalers straddling the pathway to the red tin-roofed cathedral would have caught their eye, as would the many hulks. Some of these were incorporated into jetties, whilst the three-masted *Lady Elizabeth* lay aground, but the *Phenia*, a four-masted barque of considerable size, swung to her mooring in midstream. She always lay true to the wind and was of great help when we swung past her to go alongside the old Public Jetty. She was one of the many hulks round the islands, in addition to the wrecks, to have been dismasted or severely damaged whilst rounding the Horn. Having taken refuge in the Falklands, but unable to be sufficiently repaired locally, they were denied seaworthy certificates by the Falkland Islands Company, the Lloyd's Agents, and remain there to this day.

The Falkland Islands Company, or 'the Company' as it is known locally, was formally established in 1851, but the seeds for its inception had been sown long before that. When the British abandoned Port Egmont in 1774, both they and the French abandoning Port Louis in 1767 left cattle and horses behind. With no natural predators, these animals multiplied for the next 40 or so years, and in about 1820 word reached South America that prize cattle freely roamed the camp. A turbulent period for the islands ensued, the French, the Spanish and even the Americans becoming involved and the first governor being murdered. HMS *Challenger* arrived to restore order and put their first lieutenant, Henry Smith, in place as governor, along with some seamen and a party of marines.

After four years, his successor, Lieutenant Moody, moved the seat of government to the shores of the inner harbour beyond Port William, creating Stanley. Moody also attempted to generate interest in the islands by issuing a pamphlet advertising the presence of the 40,000 wild fat cattle that could be exploited by an organisation of substance. Samuel Lafone, an English merchant in Montevideo, signed a contract in 1846 giving him the sole rights to herd and kill cattle. Despite making a profit beyond the terms of his contract, he ran badly into debt, and in 1851 his brother in London organised a rescue by the formation of a joint stock company with financiers. The Falkland Islands Company was thus created, which then received a royal charter from Queen Victoria in 1852. From then on almost every operation in the islands was either owned, managed, or reliant upon the Company. Many vessels declared irreparable and therefore as unseaworthy were prevented from continuing their voyages. Dismasting was frequent, but so too was the opening of seams during the stresses of bad weather. The vessels thus virtually impounded were used as convenient store ships for the cargoes of those vessels leaking and requiring discharge before caulking prior to resuming their voyage. Rightly or wrongly, there was even to my day a ring of monopolistic unscrupulousness about the past operations of the Company.

We went alongside the Public Jetty. There were only two other jetties, the Company Jetty, slightly enhanced from its visible origins as a hulk, leading to their warehouse, and the small wooden Government Jetty further within the harbour, adjacent to the Public Works Department. As well as our well-laden passengers, far beyond our permitted complement (for which I presume we had some sort of unlikely dispensation), we carried nine live sheep, innumerable dogs and six horses, the latter to take part in the annual races. Our reception was greater than the royal visitor was to receive a few days later; half the township had turned out to welcome their relatives and all appeared to be on the jetty at once, along with dogs running between the throng and causing mayhem, horses hitched to the frail side-rails of the jetty, and dozens of Land Rovers.

The following day, races were held on the course behind Government House, and 'our' horses did well, which pleased us, for handling them on board had not been easy. On 8 January 1957 Prince Philip arrived, and the governor welcomed him, the local defence force and the constabulary forming a guard of honour. Having been introduced and made his inspection, instead of following the highly prepared route along the front road, HRH asked to be taken straight up the hill and along the top road to see the town in its more natural state. The front road, though, in its unaccustomed whitewashed and bunting-bedecked splendour, was not forgotten, consequent upon a royal U-turn and a descent to the starting point for the tour to recommence. That same day the traditional, but postponed, Boxing Day Race Meeting was held, with most riders using a sheepskin in place of a saddle. There had customarily been a sailor's race at this meeting, and Adam and I had ridden in it the previous year. On this occasion, not only was our new captain a good horseman but also some of our newly joined Falkland Island crew were very accomplished riders. On attempting to enter, we were told that because there were so many from *Britannia* wishing to ride there would be too many entrants, so it was made a Royal Navy sailors' race. We were furious, wanting to knock spots off the navy, but HRH won anyway, having been given the best horse! That evening he awarded himself the cup at yet another ball at Government House.

The next day of the visit was a quiet day for most on board, and we managed a trip out in our motorboat. There was a reception for the townsfolk aboard *Britannia* the following day and the senior ship's officers went to it. I, however, had been invited by the Royal Yacht's Commanding Officer, Rear Admiral Abel Smith (FORY)[22] for a drink in the wardroom, for a dubious duty I had performed during the previous afternoon. Following an established practice for visiting ships, we had gone in company with some from *Britannia* to Sparrows Cove in the outer harbour, where the SS *Great Britain* lay, to collect mussels from her hull. To see Brunel's historic vessel at close hand and perhaps board her, though she was dangerously derelict, and to collect the finest mussels near Stanley was an outing not to be missed.

But the weather blew up and *Britannia*'s party, including the admiral, were disinclined to return to the royal yacht in his barge, and he asked us to transport him and his party in our

---

22 Flag Officer Royal Yachts.

workboat back across Port William, through the Narrows and up Stanley Harbour to board her. When we got to her it was blowing force eight with a good lop running the length of the harbour. Our craft had four outwardly angled lifting lugs permanently fitted to her gunwales which received hooks from a spreader hung from the crane with which to lift her. Despite what might occur in the obviously difficult conditions, and with *Britannia* making no effort to supply fenders, which we never carried, FORY insisted we put him aboard, for he had duties to attend to. So in we went, scraping a 20-foot gouge into her polished blue paintwork, and then a vertical one as we rode up and down on the swell. FORY seemed oblivious, and as he left, extended his hand and an invitation for me to visit their wardroom the following evening. This I did in calmer weather, in a boat that they sent for me!

Before sailing for South Georgia we had first to return to their homesteads those Falkland Islanders we had brought to the capital, and pick up some live sheep and pigs; it had been deemed a good idea to rear the sheep on some lower slopes of the island near the whaling stations to provide some alternative meat to that of whale for both the government party there and the resident whalers. We spent three days of interesting navigation sailing into the coves and creeks of the islands. At Port Howard a joining shackle between lengths of anchor cable parted, in similar circumstances to that of the previous season at the Argentine Islands, losing an anchor and three shackles of cable. This time the cause was spotted as being the acute angle between the shell plating and that of the hawse pipe. At our next refit the problem was rectified by a collar being fitted to the lower lip of the pipe, and lugless shackles fitted to the anchor cable. We returned to Port Howard later, after the station manager had kindly grappled for and found our chain, which we managed to raise, thus retrieving the anchor. We avoided the problem for the remainder of the season by merely ensuring the cable was always vertical when the anchor was weighed.

We finally went once more to Rincon Grande, where we loaded 500 sheep, and Salvador for a further 250 and a few pigs. Whilst our Falkland Islands crew expertly brought them aboard over a weekend, packing them into pens, Captain Norman Brown, Adam and I rode for seven hours on each of the two days rounding up a herd of steers. We got saddlesore, but enjoyed a proper taste of island life, including mutton chops and penguin eggs for breakfast. We returned to Stanley on 16 January where the crew strengthened the stowage of the sheep, for by then we had already lost seven. Few of us other than the local lads knew anything of the horrors of such a trade.

We then attracted the wrath of Neptune by embarking the Anglican Bishop of South America and the Falkland Islands on a visit to the extremities of his diocese. Seafaring folklore warns of carrying a pig and a parson at the same time; several pigs, a bishop, and a few hundred sheep spelt mayhem! The passage between the Falklands and South Georgia can be one of the worst, made permanently in the track of unhindered depressions with shifting storm-force winds and the resultant cross seas upon a gigantic swell built by their continual development. Eastgoing is usually considerably better than westbound, but not on this occasion. We progressed by either slamming so hard that long deep reverberations rang through the hull almost without pause or, when the wind shifted suddenly and the sea became confused and chaotic, rolling violently and unpredictably. Much of the time we were half-submerged. Sometimes our movement and the shipping of solid water was worse, when there was a lull in the wind and the seas still running on top of the swell were without any particular direction, thumping against us from all sides. Of concern was that when the seas slapped broadside against us, the plumes of water that rose over the vessel would go down the funnel, with disastrous consequences. We really had no ability as a small underpowered vessel in such conditions either to heave to, run before it, or even lie a-hull[23] for any length of time before being forced to attempt to resettle ourselves. We took more than a day to pass Shag Rocks, lying isolated some 90 miles west of our landfall. Surrounded by grounded bergs, which helpfully enhanced our first sighting them, and with the swell pounding both rock and ice, they were a fearful maelstrom of surging, breaking sea and spume, with a canopy of feeding birds swirling and diving from above.

For several days we took a most severe pasting, unable to take up a relatively quiet course, let alone safely maintain our preferred one. 'Storm-tossed' took on a whole new meaning. It was bad enough for us humans, but our poor sheep suffered terribly. Naturally top-heavy on frail legs, they were stowed tight together to reduce the likelihood of their falling over, but repeatedly they did so, with the crew in their midst valiantly standing them up again. The stench from their urine and excreta became appalling and invaded the whole ship. We cracked open the hatch covers as and when we could to improve the quality of their air, only for that replacing it to be damp and salt-laden, which quickly wetted their wool, making them even more unstable with the inevitable result. We lost 70 trodden under, but on landing the survivors scampered ashore at King Edward Point as if they had been on a nice day out – but more likely than not being keen to go and glad to get off the godforsaken ship. Both bishop and pigs were also hastily put ashore, and the curse hopefully lifted. Sadly we learnt some time later that His Grace, a kindly, jolly man, on a journey to visit some of his human flock up a steep, remote pass in the Andes, had exited from the rear door of a bus on the precipitous side of the road and fallen to his death. Hardly fair for a man of such goodness.

Return to the bases

The second half of this Antarctic season started properly when we left South Georgia for the South Orkney Islands late in January 1957. After that Stanley sojourn, the rude awakening of a lengthy spell of really bad weather continued. To make matters worse, we had problems with the main engines, and at best made only 8 knots, and considerably less when knocked back by headwinds. Nevertheless we went about our tasks at an extraordinary pace, bearing in mind the weather, particularly fog and snow, the uncharted nature of most of our area and our deck complement of only a master, three deck officers and eight seaman.

We arrived at Coronation Island late in the month, embarked a survey party that we had landed previously, and proceeded to our base at Signy Island. After two days replacing their stores we visited several sites, making landings for the benefit of the base surveyors and

---

23 Lie broadside onto the weather, without power or sail.

geologists, and then sailed across the top of the Weddell Sea into the Antarctic and Fridtjof Sounds. Here, on the north-eastern tip of the Antarctic Peninsula, a small hut had been established at View Point as a starting point for southern sledge journeys. A great deal of heavy ice exiting the Weddell Sea is carried through these two straits, some of it grounding, and this, together with the very irregular bottom contours, creates strong rips and whirlpools, so it was exciting navigation, steadying the ship whilst watching the echo sounder. Glancing off one heavy lump of ice, we commenced a large involuntary swing and then, caught in a whirlpool, were swung through 180 degrees before regaining control and resuming our course. At View Point, there was the advantage of an ice foot that we could go alongside to discharge the stores.

Having worked through the sounds since dawn and completed at View Point, we made Hope Bay by the same evening in bad visibility and a near-gale. We could do no more than put a boat into the base before wisely leaving that notoriously dangerous bay with its strong winds and poor holding. Northwards across the Bransfield Strait in yet more poor weather, it was a relief to see the embracing headlands of Admiralty Bay on King George Island and to run the ten miles up that bay to our base, arriving early the following day. Cargo handling was delayed on account of strong winds and breaking seas on the beach, but despite waiting for better conditions, later in the morning the lop onto the shore rolled the scow enough to upset the deals upon it down which a tractor was being discharged. Tractor and driver went into the water, but after a rub-down, neither appeared to have come to any harm; but then, with our attention distracted by extracting them, the scow went aground on a falling tide. That ended the cargo work by that method, and the stores that remained for delivery were run in by the motorboat. Adam then used it in the afternoon to lay survey marks ashore for future use, and I did likewise, but on skis with some base Fids. I retrieved the scow on a higher tide during that night and we sailed for Deception Island the next day.

On passage we diverted to a bay on Nelson Island, where we landed a survey party. The weather continued foul, and this time with a strong westerly almost directly ahead we were well delayed, taking a day to steam 60 miles. Here we anchored close to the shore in Whalers Bay, where we ran out stern lines and a hose to take on fresh water. In the evening by way of recreation we took the boat to board the wreck of the Salvesen whaler *Southern Hunter* ashore in the Bellows. At first light we sailed down the Gerlache Strait towards Cape Recluse, sounding and fixing marks ashore as best we could, but the visibility was very changeable. We took up an anchorage there for the night, having used our own chart to do so, then entertained the party from ashore. The next morning we made the short passage to Danco Island, though slowly because it was carried out entirely in fog. This then turned to wind and snow, which prevented cargo work on that and the following day until late, when we discharged a heavy generator.

The next day dawned calm with clear skies and visibility, and we took the opportunity to undertake our first proper survey of the area with the motorboat. With Adam in charge, and he and I taking the sextant angles and taking it in turns to plot, we ran continuous lines of soundings, directing the coxswain onto the required courses at right angles to the anticipated seabed contours; with a time-keeper/echo sounder recorder shouting 'Fix!' for us to act upon to take our angles as he marked the echo sounder trace, the survey went extremely well. Only icebergs occasionally obscuring our sightings of the marks interrupted the smooth chart making, whilst those same bergs caused the master and mate aboard our parent ship lying in the strong tideway to be continually harassed, having to haul in and let out the cable to speed up their passing.

During the morning of the next day, 9 February, we made a pleasant passage from Danco Island to Port Lockroy in good weather but encountered a shoal area in the middle of the Gerlache Strait, which we fixed reasonably well. The delightful harbour of Port Lockroy with its encircling ice cliff turned miserable with rain and sleet as the temperature rose, and we were glad to finish cargo work and snug down for the evening. Off again early the next day, we nipped around Cape Lancaster to Anvers Island, a passage of less than 30 miles, and there we transferred cargo to the base all day, remaining at anchor overnight with most of our cable out on one anchor in a strong wind, yet departing in fog the next morning for the Argentine Islands. At that time we were rightly beginning to surmise that in what we termed as 'good ice years', that is when there was much open water in the straits and fjords, there was a higher incidence of poor visibility, whilst in years of extensive sea ice coverage the air was marginally colder with clearer visibility. In hindsight it will seem to some to be plain common sense, but at that time we did not realise the likelihood of such micro-climates there – as indeed few yet accept the existence of the quite definite such micro-climates which surround one large iceberg or a concentrated group of them.

We remained among the scenic beauty of the Argentine Islands for three days, firstly actually going alongside the rock face of Galindez Island, where the base was situated. That was a departure from our usual running moor in mid-channel, and certainly facilitated the cargo handling. It was not quite the sheerest of rock faces, having somewhat of an awkward protrusion that in quiet seas we managed to fender ourselves against; but in worsening weather we were forced to pull off. The *John Biscoe* arrived for a short rendezvous in order to effect some transfers, and then with cargo work finished we remained to support the base personnel in their summer activities with our ship's Fids.

This left the ship's company to attend to their own jobs; whilst the engineers carried out some maintenance to the main engines, the deck crew overhauled some cargo gear, and Adam and I went off as usual in the motorboat, with a team to explore and sound out an alternative route to the base through the islands from the westward. The approach through the very narrow Meek Channel, whilst reasonably deep and clear of dangers once past Corner Rock, sited in the middle of its entrance, was often blocked by small bergs which required some nerve and imaginative handling of the ship to shift aside. Well grounded, they sometimes refused to budge, and on these occasions an outer anchorage a little distant to the west of the islands could, with great difficulty, just about be found, though none was possible to the east, nearer the base, in the deep water of the Penola Strait. The anchorage to the west was, however, amidst a myriad of small islands, and littered between them, rocks and shoals, always full of grounded bergs caught in the shallows. The area made a spectacular but daunting scene to anchor in. In the couple of days available we produced a good enough sketch survey to utilise later that voyage – and for me to use and depend upon entirely almost until I retired – with the addition of only one aid, an oil drum full of rocks and planted with a pole top-mark on an islet. Drum Rock, as it had become known, was incorporated in the full survey carried out by me some seven years later when in command.

By mid-February it was time for us to transfer our affections to the base at Ferrin Head, some 60 miles to the southward down the Grandidier Channel in which Wynne Edwards had carried out some good survey work. The *John Biscoe* already lay off the base, and our combined teams assisted in its building. Once again, this left Adam and me to survey, but with the motor launch required to support the base activities instead, and unable to sound properly without the echo sounder it carried, we set about extending our area of coverage by erecting more marks in one of our aluminium lifeboats. We achieved quite a good deal, but punctured its hull going through some brash ice and rapidly returned to the ship, several miles distant. The master thought that we had pushed our luck far enough at that point, and we started working our way northward, leaving the base personnel to live in tents for a few more weeks whilst they completed the building of their hut. Our first task up the mainland coast was to lay a depot for those sledging on the winter sea ice from the new base, but we were plagued by poor visibility, and it was essential that we knew exactly where we were placing it order for it to be found. The area at the requisite distance north was foul and totally virgin, so we had to wait for a clearance in the weather to explore it. It was quite uncanny anchoring with nothing in sight, just echoes on the radar, but not knowing whether they were bergs or islets, so nothing of reference and only vaguely knowing where we were. We sounded around the vessel at anchor in a star pattern with the motorboat to ensure that we had room to swing without dangers close enough to drag upon.

When visibility cleared a little, we shot some seal on the loose pack ice to take to Hope Bay for their dogs, but lost a couple of days awaiting proper visibility. When it came, finding an easily accessible site on an identifiable point of land rewarded our patience. On the last day of the month we weighed and reached the Argentine Islands by mid-morning. Mooring stern to, we took water, lifted personnel and departed at 1800 hours for the Anvers Island base, off which we anchored at 2200 hours, having negotiated the shoals for the first time in the dark. The next day we weighed in daylight, at 0600 hours, but despite this had to thunder astern when forced off line by ice into a shoal area.

We then progressed through the northern bases, lifting mail and making final transfers of personnel, but departed from this last flurry of activity to visit Paradise Harbour, an outstanding bay of magnificent scenery on the mainland. Here there were both Chilean and Argentinean bases, and we were to rendezvous with the *Bahía Aguirre,* an ice-strengthened transport vessel of the latter's Antarctic operation. We had been carrying an Argentine observer (it was a developing custom to carry an observer from another nation, and it became a procedure that was to be incorporated into the Antarctic Treaty a few years later), and we were to return him to their ship. We entered the anchorage simultaneously, and letting her drop anchor first, we then strapped alongside. There was none of the animosity of the previous season of protest notes being issued, and most of our officers enjoyed a drink and a meal in their wardroom. On the bridge they had some good, modern equipment, but their vessel was dirty and ill-maintained, and their manner of dress was scruffy, which all in all made us, from our much smaller vessel, feel quite superior. Adam was impressed with their charts in comparison with ours, but we later found that they were not as good as they looked. Ours were honest in showing blank spaces where there was no information, and stated when coastlines were indefinite, whilst theirs were orderly, but drawn in imaginatively and highly fictitious.

Aboard, to our surprise, was a Royal Naval captain as a British observer, and three Americans in a similar role. The captain had little time for us one and two stripers, whilst the Americans, as enthusiastic as only Americans can be when they become engrossed in a subject, talked to us endlessly of polar expeditions and exploits, and were fascinated by our survey work and progression around the region. The evening came to a sudden and explosive end with the appearance of a bishop from Buenos Aires. Whether he had just woken up or had remained seething in his cabin during our visit I know not, but he burst into the wardroom gabbling in Spanish and gesticulating. He did not approve of our presence, and not only wanted us off the ship, but our vessel to remove itself from beside theirs. He seemed to have much influence aboard and we were asked to leave, which we did; but our ship, with the complicity of their sympathetic captain, did not depart until daybreak.

For the first few days of March we lifted survey parties from Cape Reclus and Brooklyn Island, laying depots there and at Cape Anna, then visiting Danco Island for a final call, where we hoped to complete our survey. But strong winds and snow defeated us, and it was all we could do to retrieve our poles and flags from the rock cairns we had built before continuing northwards to Deception Island. A day and night there, again watering ship, for our growing complement was taking a toll on our stock. Whilst an inspection of the base by the master took place, we occupied ourselves by catching a few seal before stowing the boats and departing in the late evening. On passage to Admiralty Bay we lifted the survey party from Nelson Island, an operation undertaken in a strong onshore wind. We anchored the motorboat off the chosen landing and let a dinghy float downwind towards the shore on a line, hauling it back several times with a single person and a small load; it was a wet and weary party that we took on board and returned to Admiralty Bay that evening.

Then we crossed the Bransfield Strait to Hope Bay. During the next three days we lay there to two anchors with eight shackles out on each, yet we still dragged. Despite the motorboat grounding with some damage in the large swell, we managed to land the seal carcasses and some drums of fuel in the strong offshore wind. On 13 March we sailed through the sounds to View Point but on account of the near-gale we found it impossible to lie alongside the ice cliff, first securing our mooring lines to deadmen[24] which repeatedly pulled out, we completed our discharge by stemming the ice cliff with the engines running ahead and working the remaining cargo over the bow. We had no cause to linger there, and after lunch set sail to the southward, taking seals from isolated floes en route. We entered the Prince Gustav Channel that evening, believing we were the first to have ever navigated this channel. It had been named by Nordenskjold when he sledged through it in 1903 from the hut he occupied

---

24 Short baulks of timber dug deep into the ice; each timber having a wire strop around it for a mooring line to be secured to.

on nearby Snow Hill Island. Since that time it had been sighted or sledged through by Fids, and was always described as having shelf ice up to perhaps 150 metres in depth permanently spanning its width for almost its entire length – permanent ice similar to that of the vast, almost adjacent, Larsen Ice Shelf. The open water, which by this season had become typical, had allowed the swell to penetrate the area and crack free great flat-topped blocks of ice which the currents then carried out, allowing us to make this virgin passage.

We ran lines of soundings in the very 'steep to' and apparently danger-free channels down to Pit Point, where we found a shallow patch to anchor on for the evening and night. During that time we landed a survey party with a dog team and all their sledging paraphernalia in readiness for an attempt to reach the mainland plateau. On passage we shot seal either from the motorboat or ship's bow, and took the carcasses aboard after they had been gutted, by placing the ship alongside the floe they had lain on, and hoisting them aboard on the crane. Not a very pleasant task but good practice at ship handling, though woe betide any one of us who nudged a floe too hard, dumping men and seals into the sea. The motorboat always stood by in case of a miscue, ready for the crew to scramble into, but the catch would be irretrievable.

Our progression south the previous day had been between the islands and the mainland, so we then returned northward outside them. The desire to establish a depot and a party down the Prince Gustav Channel had given us the opportunity to carry out a unique pioneering transit of some previously un-navigated waters, and those that had been thought never to be free of permanent impenetrable ice. We had therefore definitely ventured further south than any vessel in this region. A further bonus was that the islands seen close to, Vega, Vortex, Beak and Eagle, which we had almost circumnavigated – were remarkable for their obviously apparent, strikingly differing and unusually interesting geology. We returned to Hope Bay through the tide rips and whirlpools of the sounds in yet more wind and snow, pausing at View Point to offload seal. It was becoming late in the season for this area, when advancing autumn sea ice from the Weddell Sea could pour through those sounds, so that evening we attempted to land some seal in the near-dark. But the motorboat went aground again and remained so overnight, to be retrieved the following morning, towed off by Adam in a lifeboat, he having used it and the scow to put ashore the remaining seal. Sometimes carcasses were towed ashore on a strop astern of the motorboat, but mostly they would be put on a platform of deals atop the scow, then manhauled or towed by tractor to a seal pile. With the blood and stench, which took months to remove from our clothes, it was never a pleasant task.

Sailing at 1400 hours we arrived at Admiralty Bay in the early hours of 17 March. Having seen little ice as we crossed had the Bransfield Strait, and the wind remaining westerly whereas the bay was open to the south, it was judged that no pack would enter the deep indentation of the bay, so we planned to remain there for three days to complete our survey of the anchorage. Care had to be taken here towards the end of the season that pack ice did not trap a vessel in the anchorages 12 miles up at the head of the bay, with the chance that she would remain beset over the winter. By the evening of 19 March we had completed our survey in very poor conditions of wind and snow, with the spray sticking to our sextants in the cold as the boat was rolled and slapped by the short sea. Sailing for Stanley, for it was certainly time to be withdrawing, we were momentarily glad to depart these chilly waters, then saddened that another season down south was over.

Aboard we had become a close-knit team. The master, in our humble eyes, was very competent and adventurous but aware of the limitations of his vessel, brave in his navigation but cautious in ice. He was friendly but authoritative, and seemed in the wardroom to be more like an older brother than the captain. We deck officers had naturally slipped into a mutually satisfactory mode whereby the chief officer, Tom Flack, mostly kept ship whilst we two younger mates carried out nearly all the boat and shore work. Adam was navigator, doing the passage planning when required, keeping the navigational log and preparing the hydrographic notes. I kept the meteorological records and tried my hand at forecasting, was responsible for the boats and was general dogsbody to my three seniors. Often we would work all day and then steam overnight, then the master would take the bridge for a few hours whilst we got some sleep. When we attempted running surveys on passage we would all, captain and three deck officers, man the bridge, taking simultaneous sextant angles and bearings, manning the echo sounder, plotting, drawing in the coastlines from the radar, and advising the new course to be steered to maintain a sounding line, whilst watching where we were going! The engineers cheerfully, though sometimes with difficulty, kept the wheels turning down below as well as maintaining the steering gear. Likewise the deck machinery, particularly the crane, windlass, and the boat engines, all in constant use; on many occasions maintenance was a difficult job on account of the freezing temperatures, so when we ventured far from the ship an engineer would accompany us. The radio officer, in addition to his normal duties and watches, maintained contact with the Falkland Islands office, all the bases and shore parties, and the boats when away, and whilst south was recognised to put in at least twice his normal hours. Unfortunately our experience with boat radios was poor because their batteries failed in the cold, which often caused concern. The chief steward and cooks, as on most ships, were temperamental, but kept us very well fed, sometimes overcoming the most appalling movement to work in the galley. Everyone worked hard when necessary, and we were all content to lie up when it was not possible to do so. The UK component of the crew eventually appreciated the difference between this ship and a normal merchant vessel, and although that did take time, it was helped by the Falkland Islanders aboard being accustomed to that way of life. In all, down south we were a happy, busy ship.

We arrived in Stanley on 23 March, aware that in the absence of any link between the Falklands and Montevideo – because the Falkland Islands Company's ship *Fitzroy* had sailed to the UK prior to its new vessel, the *Darwin,* arriving – it was we who were destined to fetch the new Governor of the Falkland Islands from Montevideo. What we had not realised was that the round voyage was not scheduled until some three weeks later. Overnight we changed from being a hard-working polar research and survey ship to an adjunct of the Colonial Office. No one in authority appeared to recognise the importance of utilising such a major asset as a ship for as much time as possible. When we could perhaps have been surveying – for the charts of the Islands were mostly more than 100 years old with only minor modern corrections – we lay alongside the Public Jetty. We closed the ship's Articles for the current voyage, and re-signed for the vessel's third Antarctic season on 1 April 1957, although in practice we would not embark upon it until after sailing home to, and then departing from, Southampton in the northern autumn of that year. Perhaps we should have noted that date of re-signing in the light of what was to happen on that third voyage, for fate certainly made fools of us.

Eventually we sailed to collect Edwin Arrowsmith, the new governor, from Montevideo, but had two seamen desert there and returned to Stanley shorthanded. We signed on two new Falkland Islands crewmembers and at long last headed for home, again via Montevideo, where we embarked three displaced British seamen. Ships of the UK are obliged by law to afford such seamen passage to work their way home, they having either been left behind sick or missed their ship. Knowing that as always, we were bound to run out of fresh water before reaching home, we laid off courses to replenish at the Cape Verde Islands. In the event, falling short even before the Equator, and having a sick crewmember whom it was better to land, we called at Recife. A mistake, for the Brazilian customs and immigration took two days to 'clear the ship in', by which time our patient had recovered, and the watering had only taken about two hours.

Captain Brown was uneasy about the performance of our steering gear and took the opportunity whilst in port to call in the local Lloyd's surveyor. Obviously it had been given some rough treatment, backing and filling in ice, but from tests alongside nothing was found amiss.

We paused yet again for more fresh water, this time at Dakar in West Africa, before anchoring off Bournemouth to await a planned arrival. We docked at Southampton to a fabulous reception during the forenoon of 5 June, moving on the afternoon tide to Thornycroft's yard at Northam for our refit

Standing by refit – gaining a certificate but losing a friend

The summer of 1957 was a very busy one for me. Whilst the ship underwent her refit, we three deck officers shared equally the period of overseeing the work in the shipyard. Adam and I also sailed in our yacht *Marishka*, but more importantly I sat the examination for my first mate's certificate. The sea schools and examination centres closing for the summer recess, and a protracted discussion with the examiners regarding my sea time, did not make this easy. Two questions were raised: first, did the three months spent in the shipyard, albeit signed on the continuous Articles of the ship, justifiably count as sea time towards that required for sitting the exam? Secondly, did our voyages to the Antarctic, whilst undoubtedly 'foreign going', accord with the examiners' ideas of a voyage to foreign ports, dealing with commercial matters? Fortunately my arguments carried the day and set the precedent for those that followed me.

Passing the exam a couple of weeks before we sailed, I rejoined the ship with all the crew to prepare for another departure, only to be very surprised at the news I received. Adam was to leave. His father saw little future for him with the Survey, and whilst on a trip to Canada had enquired regarding a position there for him with the Department of Mines and Surveys. The master agreed to let him go, as long as a new deck officer could be found. On sailing day, 1 October 1957 he, having only signed off that day, watched from the quay as we moved away with a new man engaged as third officer, I being promoted to second.

5 RRS *Shackleton*: Third Voyage

Near-disaster and far-reaching consequences

T he third voyage started reasonably well, even though our new third officer had neither any enthusiasm for the south, nor any interest in learning to survey, nor the energy of Adam. Not so surprising, for he had been the only certificated officer available from the shipping office pool the day before we sailed.

We made the usual calls at Montevideo and Stanley before sailing south, on this occasion briefly to the peninsula before setting an easterly course from its northern tip, south of Elephant and Clarence Islands, to Signy Island station in the South Orkney group, a passage across a stretch of sea where there was always a high possibility of meeting much ice in the spring as the pack broke up in the Weddell Sea and moved north. We met quite heavy multi-year ice covering seven-tenths of the surface of the sea, which forced us north of our direct course towards Elephant Island, its coronet of 12 peaks clearly visible without a vestige of cloud in some unusually fine weather. Cape Valentine, where Shackleton had landed after losing *Endurance*, beset and crushed in the Weddell Sea in 1911, and Point Wild, near where he had set up camp, were not visible, being on the far side of the island. Shackleton and all the members of his expedition had made their way by hauling their 24-foot boats over pack ice, then sailing them in loose ice and in open water to a cove of relative safety. From the camp he established nearby, he later embarked with five men on his remarkable 800-mile boat journey to South Georgia, eventually mounting a relief expedition and returning to Elephant Island for the rest of his men. Shackleton is the hero of most involved in the southern continent, so we doffed our caps as we slowly wove our way through the ice past this island, so historic in the annals of polar exploration.

Very soon, though, the next island in the group, Clarence, took our attention. With Mount Bowles rising to 6,000 feet it is, as Dr Bruce of the 1903/04 Scottish Expedition described it, 'wild and majestic'. It is also beautiful and dramatic, with very precipitous dark cliffs capped by sheer sparkling ice and snow slopes, which we viewed in all their sunlit glory as we crept eastward. The barometer continued to fall, but there was not a breath of wind or a cloud in the sky. The ice began to tighten around us, and any small leads, or pools of open water between floes, closed under the pressure that visibly built. Floes ground against each other and existing pressure ridges, a few feet in height at their edges, toppled as far as the horizon as the mass beneath them was disturbed, only for more ridges to appear as new rafting[25] took place. The

---

25 Ice floes in a field of pack sliding on top of each other when under pressure.

northerly stream of ice was, on exiting the Weddell Sea, being jammed against the two major islands of the Orkneys, close to our northward. We became beset for several days, without a zephyr of wind, yet extraordinarily with the barograph trace off the bottom of the chart, at around 938 millibars.

We were experiencing a rare Antarctic phenomenon, an enormous static polar low. Very occasionally a number of areas of low atmospheric pressure over the continent become enjoined to create one vast cyclonic area within which there is little or no barometric gradient and therefore no wind. Our little ship was not built for such forces of nature, yet amazingly we suffered only some minor distortion of plates and frames. When not laboriously collecting blocks of ice from the pack ice, we sunbathed, shirts off, in temperatures well below zero. As usual we had run short of water, and boiling ice down on deck over diesel oil fires provided some, it being mostly composed of fresh water, though the resultant water tasted of diesel and was distinctly brackish.

On the ninth day our calm sojourn amidst the tortuous ice ended as a northerly breeze set in and the pack loosened, enabling a resumption in our passage to Signy Island Base. This we reached and, relieved, then departed eastwards, heading for the straits between Coronation and Laurie islands in fairly open pack ice. As we left the straits, working our way northward and westward to clear the main body of ice, it became more open as it escaped the confines of the island group and spread freely into the Scotia Sea, loose enough for it to be easily navigable at a reasonable speed. As second officer I kept the middle watch and was due to take over at midnight from the master. At 2350 hours on 29 November, having been called, I had dressed and was putting on my fur boots when the ship gave a violent lurch, enough to make me fall. I reached the bridge just moments after the impact, keen to work some ice, only to find that we were almost certainly holed.

At this time, beneath the surface of apparent equanimity between the Fids aboard and the ship's company, there was a degree of disharmony; this was caused, I believe, by the cargo work being hard and intense during the initial part of this voyage and this particular group of Fids being none too willing labourers, and then that period followed by a spell of idleness and possibly boredom whilst being beset, creating an environment in which grievances could fester. Such harbouring of discontent rarely occurred, but did sometimes surface during voyages. Perhaps, in hindsight, criticism of the organisation could be levelled for not making the Fids aware, when they were recruited, of the duties required of them such as assisting to maintain the ship, and work cargo. Normally the successive changes of personnel on board occurring whilst at the bases, the arrival at an individual's destination, the interest of the scenery and wildlife and the anticipation of their years on base would overcome any dissatisfaction. Further, most of the time the Fids were appreciative of our safe transport of them to their destinations, and our pushing far beyond the boundaries that would normally have been imposed by the weather, ice and uncharted waters to further their aspirations.

We never lost a man in these endeavours, let alone a ship – but that night we nearly lost both. After this exceptional instance a few young scientists sowed seeds of discontent amongst a number of their fellow Fids by implying that they were intellectually superior to the ship's officers and that they, the scientists, could have organised our affairs and run the ship better than we could, particularly in the handling of this incident; in their eyes we were little more than taxi drivers. They failed to recognise, let alone appreciate, the professional judgements we exercised, and indeed had no comprehension of our work environment and that it was so different from the norm. They also failed to understand the need for, and resented the authority of the master and his officers, and the discipline imposed. During the narrative that follows, which is compiled from my notes written in South Georgia about one week after the event, I now take the opportunity not only to relate the events as they unfolded, but to rebuff some of those criticisms. For those who do not know the detail of this episode it provides a further dimension to the story.

The third officer, whose watch was just ending, was totally inexperienced in ice, and the master, Norman Brown, had been on the bridge since 2000 hours handling the vessel. Weaving between floes, he caught the ship's side just forward of amidships on one that contained a hard glacial core with little softer snow cushioning; I had felt that lurch. One account by a Fid, many years later, stated that there was no officer on the bridge that night. Never during my time with the survey was a bridge left without a watch-keeping officer whilst under way at sea – let alone in ice, which would have been suicidal – and it was quite ridiculous to suggest such. From the very first departure from Denmark, we consistently had steering problems, inasmuch as we usually carried five degrees of helm[26] one way or another, depending on the combination of the ship's state of loading and from which quarter wind and sea came. She also had a tendency for her bow to fly off into the wind. The opinion of all those who understood the matter was that Brown had probably misjudged a gap between floes, possibly with more way on than he would have preferred. Our method of progress in this type of loose pack would be to balance keeping good way on, at about 4 knots, to facilitate handling but then slow to provide just steerage way through narrow gaps, sometimes enhancing that by short thrusts of power to assist a swing or maintain direction. Some hard contacts, which occasionally resulted in indentations to the hull, were unavoidable, and were acceptable hazards of such passage-making.

As I arrived on the bridge Brown told me to have the tanks and bilges sounded[27] in the vicinity of the collision. Tom Flack also came to the bridge and aired concerns regarding the flooding of the main hold. Not long after I had called the bosun's mate, who sounded the fresh water and ballast tanks daily – for we carried no carpenter who on most ships undertook this task – he came to the bridge with the bad news that six feet of water had already entered the main hold. He would possibly have told a few souls of his findings on his way there, for he would have passed the Fiddery (Fids' mess) en route to the bridge as he climbed the inner companionways. He did not, however, discover water gushing in somewhere and report it of his own volition, as was reported later. The master then consulted Tom Flack as to whether the remaining intact spaces would provide sufficient buoyancy to keep the ship afloat with such an amount of water already in our main space. Tom, with his intimate knowledge of our stability, advised that we would not stay afloat for long. Consequently Brown deemed that the

---

26 Having to steer five degrees into the wind to maintain a course.

27 Plumbed to detect and quantify the amount of liquid in a tank or bilge.

best objective was to attempt to beach the vessel. We knew of a reasonably satisfactory site at Uruguay Cove on Laurie Island, which also happened to be one of our nearest points of land, although still over 50 miles distant, and attempted to head there as we transmitted an SOS. The pack, however, became more dense in that direction and not only impeded sufficient progress, for time was of the essence, but greatly hindered the rigging of a collision mat overside to stem the ingress of water. The chief officer took overall charge of everything about the deck, the third officer was given the task of rigging a collision mat, and I was assigned to the main hold to reach the damaged area and make it as watertight as possible from within.

We opened up the hatch to the main hold, where the damage was thought to be, and that to the small fore hold on the forecastle. From both holds we jettisoned cargo by crane and by hand in order to lighten the ship, to bring the hole nearer the surface, and to facilitate working on it from the outside and reduce the pressure of inflowing water. Just in time, drums of cement fondue which were destined to form the foundations of a new base hut were prevented from being discharged, for it was hoped that this could be used to build a cement box over the damaged area. It was frantic work – but not a chaotic shambles run by idiots, as one unkindly Fid wrote. The method used to place the mats over the hole was to attach them to ropes which were passed under the bow, then worked aft beneath the hull whilst keeping the mat itself clear of the water to minimise drag until it was vertically over the damage, when it was then eased down to cover it. As we were repeatedly losing mats when they were removed by passing ice but still not making enough progress towards the cove, it was decided to abandon any attempt to reach the shore and put every effort into plugging the hole from the inside, whilst a mat was hopefully kept in place outside with the ship stationary. My team and I had by this time reached the 'tween deck hatchboards, which we lifted and then entered the lower hold to find the flood water only about one foot beneath the deckhead.[28] We entered the freezing water and began to discharge the lower hold cargo, gradually working our way towards the ship's side beneath the beams and deck plating above. Several times we were driven out by the rising water denying us those few inches of air space to breathe in. The pumps mostly kept pace with the ingress of water, and even at times gained on it, but then unaccountably lost the battle altogether, for the water to rise above and flood across the then partly empty 'tween deck.

Until this time, with the loss of buoyancy due to the flooded hold, the vessel had taken on a sluggish but not undue roll in the slight seaway, but then it became apparent that she had developed a list to one side, which quite clearly became a loll, which transferred periodically from one side to the other as the seaway dictated. With the flooding, albeit only to a few inches in depth, over the entire 'tween deck, we were experiencing an unwelcome classic example of the free-surface effect of water, whereby the stability of a vessel is so detrimentally affected; this is well known to mariners from their studies, but rarely experienced in practice. The dynamics of that mass of water, alternately weighing down on each extremity of beam, increasing each roll, and that ending with an agonising loll, sometimes being pushed further over by the sea before returning through the upright to repeat the cycle in the other direction, could easily have become the factor that capsized us.

---

28 Ceiling of a space within a ship.

A vessel in this condition cannot float upright. As a ship heels over, her centre of buoyancy moves towards the low side. If the centre of buoyancy shifts sufficiently it can align with the common centre of gravity of the ship and the internal seawater; equilibrium will be achieved and the ship will float with a loll. But if the centre of buoyancy does not move out sufficiently she will capsize. With our inability to float upright and reduced stability to counter rolling, it was fortunate that at this time there was not a great deal of sea running.

At this point, with the possibility of capsize in mind, the chief officer decided to lower the lifeboats into the water to lie alongside, manned by their coxswains. But this was not easily achieved. The boat falls, the wires with which the boats were lowered, ran through sheaves along the deck. For some days snow had accumulated over the little-trodden boat deck, the upper layers sometimes turning to sludge by day, but to ice by night, and this formed a layer several inches thick, particularly where snow had been caught around the sheaves and wire falls. This had to be axed off before the wires would run. Again, later, it was suggested that neither the boat falls nor sheaves were oiled and that moving parts were clogged with paint, yet I had lowered both boats for use at Signy a short while previously. A further aggravating criticism of our behaviour at this time was that initially the crew were all wearing lifejackets, whilst the Fids had none. The simple explanation of this was that on the voyages during those early years it was customary for the crew to keep their lifejackets in their cabins and they had brought them with them, as instructed, when called on deck, whereas the Fids and other visitors' (officially supernumeraries') lifejackets were stored in large teak lockers on the boat deck from which they should have collected them as instructed at boat drills, but had not.

In the hold we continued to gain and then lose our position with the flood water; then it was decided at a point of some success to induce a permanent list to starboard to obviate the likelihood of any further free-surface-effect rolling were the water to rise above the 'tween deck again, and also to bring the damage nearer the surface in order to work on it more easily, or indeed at all, from the outside. This was achieved by lifting the motorboat from its housing within the scow on deck, and hanging it off the crane at the extremity of its outreach. It was certainly a major gamble, but it was thought that if the water rose once more to flood the 'tween deck, the position of the boat could be adjusted to compensate, or it could simply be put in the water or rehoused on deck.

The collision mats were made up from any canvas and mattresses we had, but the wind although light, was enough to give us some drift, and that movement was enough to force more water under our collision mats and into the vessel. Ice alongside also became a major problem. Despite the master's efforts to balance the position, restricting our movement through the water to reduce the lifting of the mats from against the hull, and keeping ice away from our damaged side whilst safely maintaining our list, we repeatedly lost our stock of canvas and mattresses. Because of the swell height we knew that the ice edge was not too far away, and although within the pack ice the mats were consequently being torn off, the ice dampened the swell and it was thought safer from the point of view of stability to remain in it rather than make for the open sea. Therefore, whilst the engine room remained unflooded, working this compromise position was accepted and maintained against a northerly drift towards open water. It also suited any possible evacuation into the lifeboats because they could be hauled out onto larger floes, and personnel would thus be better protected from the weather, and therefore safer than in the open sea.

Meanwhile, the jettisoning of cargo continued by hand, the motorboat being hung from the crane, although large items of not too much weight were moved from the lower hold to the almost empty 'tween deck. As we progressed down into the hold it became apparent that there was no correlation between the water level rising and falling, which it continued to do, and the ingress of water from the damage that we were now very near. We therefore adjusted our efforts to make a two-pronged attack, one to continue to reach the damage, but a second, more ambitious, to reach and clear the bilges that were still well beneath us, for they were obviously not being pumped satisfactorily. We worked on and off as the height of water allowed, with sometimes just a headroom of air between its surface and the deckhead, and staying in the water for as long as we could bear – it being minus 2 degrees centigrade – or beneath it for as long as we could hold our breath. When the pumps failed to deal with the influx of water, it rose and forced us up into the 'tween deck. Here at these times we were able to strip, rub down with towels and have a hot drink all supplied by an untiring mess boy running to and fro with all that was necessary, before putting on some clothes again, mostly wet, and readying ourselves to plunge again when the time was right.

Eventually, through a combination of our efforts, the overside collision mats and luck that at times the bilge pumps operated, we managed to reach and clear the strum boxes[29] in which the bilge suction pipes were located. We found that they were clogged with a mixture of labels washed off from tinned food amongst the stores destined for the bases, and soft, internally, formed ice. They were easily cleared and were kept clear, underlining how upon such simple things the day could have been won or lost. This done, the water level immediately dropped, which meant we now only stood in water; but we were faced by an incoming torrent as we attempted to clear an access to the hole itself. The gash in the ship's single-plated hull was no more than 36 inches in length and at most 12 inches from top to bottom, but the damage to adjacent frames and plating was severe and much more extensive. Reaching the ship's side we first put in place some framework across the hole for the external collision mat to rest against. Unfortunately the mat split and water exploded all over us, but we gained some respite with more coffee whilst those on deck rigged yet another mat. Once that was in place, we sandbagged the area and built a box to strap against the ship's side to take the quick-drying cement. A Fids carpenter, drawing upon materials destined for him to build a new base, led the construction of the box, and the third engineer brought a chain tackle from the engine room with which to bowse[30] the box onto the ship's side, securing it fore and aft to lightening[31] holes in the nearby frames. Then a sand and cement mixture was poured into the box as sandbags were removed. Several times over a few hours, success was thought to have been

---

29 Perforated metal enclosures to keep bilge suctions clear of debris.

30 Nautical term meaning to tighten against.

31 Holes cut into steel components such as plates and frames to reduce their weight without compromising their strength.

achieved only for a leak to appear, then a stream, then a gush, followed by an explosion of water. We would then be made aware that the current mat had been displaced, and on deck and below we started all over again.

After a few minutes short of 24 hours since I had gone to the bridge for my watch, my team and I emerged up on deck from the hold, the cement box finally holding secure, and the hold almost dry. My companions down there were almost entirely Fids; having only eight deck hands, who were manning the boats, the crane, the gantry and the wheel on the bridge, this left few to be elsewhere. These Fids, as did the others on board despite the criticisms I have levelled at some for their subsequent remarks, worked tirelessly to save all our lives throughout the long episode, and in most cases, as in the building of the cement box, we would have been greatly compromised without them.

Once up on deck, I found that there had been various concerns on the bridge, leading to some lifesaving stores being placed on a floe tethered alongside in readiness for abandoning ship. One such concern was that water had been found rising in the store spaces that lay immediately abaft the hold, the other side of a watertight bulkhead, yet there was no sign of any damage to the ship's hull abreast that area. It had peaked there at about two feet in depth, entering the wire-meshed food lockers but was kept out of the refrigerated compartments by their sealed doors. It had subsided, however, when the main damage was plugged. A member of the engine room staff had been stationed there to report on it and once when visited he had been found sitting contentedly on a box, feet in the water, opening all the sodden cereal packets to collect the plastic toys contained in them. As amusing as this was, the water in these compartments was not, as it constituted a major threat to our overall buoyancy. We could only guess that there must have been a leak between the main hold and the stores compartments somewhere at the watertight bulkhead.

Having retrieved the emergency stores from the ice floe and stowed the lifeboats, we moved delicately north out of the ice with the motorboat overboard, but in this instance bowsed against the foredeck bulwark to give a degree of heel to raise the damage higher in the water. On reaching open water we stowed the motorboat back within the scow on deck and set a course at slow speed for South Georgia, some 500 miles away, where there were three whaling stations with repair facilities, one of them, Salvesens at Stromness on the north-east coast of the island, having a dry dock. We had broadcast our plight in the form of an SOS, which extraordinarily, by a freak of skipping radio waves, had been picked up first by a vessel in the North Atlantic but not heard at all in the Falklands/South Georgia area either on land or at sea. That vessel, on receipt of our message, relayed it to the Salvesen Whaling Company in Leith, Scotland, who passed it to Cape Town and to their whaling station at Leith, South Georgia.

The latter then directed it to their nearest whale catcher, the *Southern Lily*, to find us and stand by. A day after we entered open water she arrived, and after a brief radio conversation, which must have satisfied her as to our seaworthiness, and no doubt influenced by her desire to resume whaling, she left – though not before coming up astern and throwing us a heaving line to which they had attached a bucket containing their outgoing mail for delivery to South Georgia; undeniably, but remarkably, a vote of confidence in our ability to get there.

Little did they realise the risk they were taking with their love letters. No sooner had we parted company, now steaming in an upright position with no collision mats over side – they being ineffective at any speed – than the cement box gave way. Water poured into the hold again and almost immediately appeared in the adjacent storeroom indicating a leak in the bulkhead at a very low level. In the seaway of the open water we began to roll alarmingly, the floodwater rushing from side to side, making it extremely difficult to work within the hold. We hove to, the master quietening her down as much as possible, and turned the motorboat out again to give a list to raise the damage; and whenever the pumps gained on the floodwaters we set to work on repairing the cement box. This cycle of stowing the boat, steaming, heaving to, putting the boat overside and yet again repairing the box, then resuming the passage, took place four or five times.

After almost two more days of such progression, help arrived in the form of HMS *Protector*. She was the Falkland Islands guard ship, an old netlayer fitted with a hangar and carrying a helicopter. She was visiting South Georgia with the new governor, Sir Edwin Arrowsmith, aboard on his first visit to the Dependencies, when she was alerted of our predicament. It was said later that she did not hear our distress call, her reception being blanketed by the mountains surrounding her anchorage at South Georgia; and I am of the opinion that receiving the news from Salvesens, and in the context of their catcher going to our aid, she had expected merely to escort us to South Georgia, believing that the *Southern Lily* would not have left us in other than a seaworthy state. She was therefore quite unprepared to provide any practical help. As it was, we were struggling to maintain the status quo, and any assistance was more than welcome, if not imperative. *Protector* had amongst her equipment a portable salvage pump and a pneumatic Cox's gun that enabled bolts to be fired. We carried a variety in size and thickness of steel plates, intended for any repairs undertaken in ports. There was some radio dialogue as she hove into sight, which resulted in an offer for a commissioned shipwright, George, and a petty officer shipwright assistant to come across to us with the pump and the gun.

But to our amazement *Protector* would neither put a boat down nor use her helicopter to deliver them. At the time of her arrival we were wallowing, settled well in the water with both the hold and stores compartment flooded, causing us to loll dangerously in the seaway, and with the motorboat consequently stowed, an obviously delicate and worrying predicament, as observed from a distance – which I did when, outrageously, we were forced to put our boat into the water and test our ability not to capsize in order to execute the transfers. I took the boat in the then moderate sea conditions to their ship's side. The short passage was of no consequence, but attempting to load the heavy pump into our open boat with no room within to evade it overhead as we rose and fell on the swell was daunting. We made several attempts to land it, but what was frustrating and making it more dangerous was finding that the lifting strop used around the pump was shackled to their crane wire and could not be released instantly as required. After much shouting at them, I took the boat off their ship's side whilst they fitted a quick-release hook; then I took her in again and fortunately at first attempt was able to land the pump whilst the shipwrights, already in their wetsuits, leapt aboard. Once in the *Shackleton's* hold, the pump easily dealt with the extraction of water. The two shipwrights with their training in damage control were invaluable, and in addition to exercising that knowledge, they relieved our officers, some of who had been on their feet for most of 72 hours. Tom Flack was so exhausted that sitting on the wardroom deck, he fell asleep there for several hours.

Almost totally watertight and with stability regained, our stalwart shipwrights keeping an eye on our damage, and with *Protector* in attendance, we began to steam slowly northward. But then after only the first few miles the cement box exploded, and it was decided that despite the moderate sea conditions an attempt would be made to fix a plate over the damage from the outside. The motorboat was lifted from its chocks and hung overside to give a heel as we rigged a stage at the water level from which the shipwright could work, at about two feet beneath the surface, to secure a plate. A rope was passed beneath the ship from the starboard side to the stage on the port side to prevent it bouncing on the swell, while Brown sought some relief from the swell, rather unsuccessfully, behind an iceberg. George, with his head frequently beneath the water, then fired bolts into the ship's plating near the damage, leaving shanks externally to which nuts could be attached. He then came back aboard for some respite before offering a large steel plate up to the bolts several times to locate where holes had to be made, all taking some hours of cold, wet work. The plate was then taken to the engine room to be drilled.

George took a breather on deck whenever possible, but unknown to several grateful crew and Fids was the fact that they had each been plying him with a tot of rum when he did so. Finally, being helped back inboard over the bulwark onto the foredeck, and having already taken a voluntary swim away from the ship on his lifeline, he was found mysteriously to have lost his sea legs! His assistant completed the operation, fixing the plate against the ship's side with a fire hose filled with oakum[32] between to act as packing. George wished, though, to show everybody how effectively his Cox's gun operated; before we could turn off the compressed air to stop him, he had made the fashion plate rising from the foredeck to the forecastle resemble a colander. The general mirth that ran around the decks was, I suspect, also an indication of the relief that everyone felt at the hole being effectively dealt with, the closeness to losing our ship only really dawning after the event had passed.

As we steamed north at about eight knots it was still necessary to have a cement box backing the damage because the repair was found to be not entirely watertight. Continuous monitoring of it was required, and to this end we posted two hole watchers; they took turns to sit in the near-empty hold to observe any change in the ingress of water. The two gentlemen allotted to this task were Professor Jim Cragg, a biologist from Durham University who was with us to view his discipline and form the programme for his researchers for next few years, and Llewellyn Chanter, a senior diplomatic correspondent from the *Daily Telegraph*, who at the invitation of the Survey was to report on our work and comment on the political aspects. They were both delightful companions in our small wardroom, intelligent, knowledgeable and entertaining, as were most of the visitors we carried, that being one of the bonuses of our work. Jim Cragg was dedicated to his subject, had little time for superficial matters

---

32 Tarred fibre used to fill gaps in ships' hulls.

or pretentiousness petty authority, and could appear abrupt. On arrival in Stanley he had upset the hierarchy when, having declined an invitation to the usual welcome cocktail party at Government House, he chose at that very time, in full view of all those attending, to scrabble about on the foreshore in his sea-boots and southwester looking for marine bugs. We sympathised with his not wanting to waste precious time with small talk when a foreign foreshore could be explored, and his behaviour endeared him to us. Lew Chanter did the same, for different reasons. Extremely easy to get along with, he was perceptive and articulate. He was now sitting on a world scoop, and had written an account as the episode evolved. Because our radio room was dedicated to emergency traffic he had arranged with the captain of *Protector* that he would pass his copy to them by radiotelephone for onward transmission. He had observed *Protector*'s poor showing, and shared our indignation and displeasure over their apparent disregard to our plight, and their minimal co-operation. Whilst giving praise to the shipwrights, he damned their overall performance. So the captain of *Protector* refused to send the copy unless it was altered. Chanter removed his derogatory truths, the message was sent and his account made the top left of the *Daily Telegraph*'s front page.

Without further problems, we reached Stromness, Salvesen's engineering and ship repair base on South Georgia, where they had a small floating dock for whale catchers. But we were too large and too heavy for it; the nearest dry dock of sufficient size to take us properly was near Valparaiso, westward across the Drake Passage, around Cape Horn and 1,400 miles up the coast of Chile. But in the isolation of South Georgia the whaling managers had a philosophy of solving problems themselves, and at Stromness they agreed to attempt to raise our forepart in the floating dock whilst our stern remained afloat. This was achieved, and at a very unusual angle of repose for a few days, our damaged hull was completely rebuilt. In addition, hose-testing found a poorly executed attachment at its foot of the watertight bulkhead between the main hold and the stores compartments, and this was sealed. That single, apparently minor, faulty piece of workmanship would have resulted in sufficient loss of buoyancy and free-surface water for the ship to have been lost. It underlined the need at sea, especially in our waters, to be ultra-particular in all matters that impinge upon safety. At this time the seeds were sown within me to be uncompromising on all matters concerning safety, not only whilst afloat but even in the shipyard, where my requirements often went far beyond those of marine surveyors.

We held a debriefing session amongst ourselves whilst in dock, at which I took notes, and my comments on the happenings about the ship during the incident, especially during my time in the hold, stem from that meeting and those notes. I also began to compile a notebook within which I enumerated all the deficiencies of our present ships that I knew of, together with all the desirables and requirements that should be built into any new tonnage. I kept it going throughout my career and it served me well. We learnt a good deal from that incident, much of which was rectified at the next refit. Our headquarters, and a plethora of advisors who appeared on the scene later, suggested that our lifeboats should have aboard extra rations, primus stoves, snowshoes and skis; the list was endless, enough to leave no room for any occupants! We did, however, follow their advice on the carriage of survival suits, which could be utilised when abandoning ship but were also strong enough for working underwater in the hold. In hindsight, a great deal of that which was necessary should have been obvious and put in place before any ship ever ventured to the Antarctic. I have had it said to me that we were naïve to have sailed in such a poorly designed and equipped ship – but even those with experience of the conditions we should meet, such as Johnston and Brown, let alone the newcomers, were of the expedition mentality, making the best of the tools we had. Also, in fairness to ourselves, no ships' surveyors or naval architects had ever suggested putting in hand before that voyage any of the alterations and additions, such as wing tanks in way of the hold and double-bottom tanks beneath the engine room, that were installed subsequent to the holing. From early on, the Crown Agents had been commissioned to oversee refits, and they had no knowledge of polar seafaring, and our very small organisation of only three men and three women in our head office dealt mostly with personnel, stores and financial matters, with very little input into anything related to the ships. We were in a transitional stage, growing from a temporary expedition to an ongoing government body.

The resident administrator at South Georgia held a preliminary enquiry into the accident. Without any nautical advisors and he having no nautical background, it was inconclusive; but Brown raised the question of our having defective steering gear, and that it had been looked at during the previous homeward voyage but found satisfactory, and that was noted. After this we sailed for Port Stanley where Tom Flack ran up some calculations, later verified by the Board of Trade, that spelt disaster. He calculated that during our flooded state we had had a negative metacentric height that would have induced a loll of about 45 degrees. These findings were sent to Captain Freaker and Mr Cox in Southampton, who had them verified. What had actually kept us afloat was never discovered.

We returned to Stanley to undertake an audit of the cargo that had survived, in the expectation that replacement stores could be shipped to us there in time to be taken south before the end of the season. To utilise the time of this enforced delay we made our way back to South Georgia to carry out some surveys. Tom Flack and I surveyed Rosita Harbour on its north-west coast over a period of a few weeks, and the resultant chart became my first to be published by the hydrographer. FIDS costs were partly met by the Falkland Island government, which in turn derived much income from its taxation of the whaling industry. A fair and sensible return for its financial contribution had been suggested, whereby our surveying capability would be utilised when we were not working south, immediately before or after our southern season, to chart some of the anchorages favoured by the whale catchers, which they were unsure of being clear of dangers.

The human toll

Returning to Stanley from Georgia to load the goods destined for the bases, we experienced a drama which was to set in motion a train of events that would affect us for some years. The master, Norman Brown, deserved much credit for his calm handling of affairs immediately after we had been holed. Throughout his time aboard he was confident and in my opinion at that time competent, with a natural ability and qualities of leadership. Being comparatively young, his style was to treat us as colleagues rather than subordinates. He was likeable and friendly, and ran a happy ship.

To this day I can only surmise upon what brought about his sudden mental collapse and resultant extraordinary behaviour. Possibly there was an underlying strain in place from his command south, hidden from us by his apparent bonhomie. Was the question of the ship's steering of concern? Perhaps his error in holing the ship off the South Orkneys played on his mind, and maybe he feared returning south with the occurrence of an even worse incident. Possibly, after the easing of tension whilst in port then being confronted with the realism of sailing south again, on the day before we were due to leave he was discovered smashing the bridge equipment with a fire axe. His venom appeared to have been concentrated initially on the telephones, they perhaps epitomising the many problems facing a master in those waters. Unfortunately, between being found on the bridge and taken ashore, he sat in his cabin and drank, leading many to believe that he was drunk when discovered on his rampage, which he was not.

The whole ship's company underwent a particularly unpleasant enquiry over a period of several days, chaired by the colonial secretary, the officer immediately beneath the governor. Brown, serving as chief officer in *John Biscoe* prior to his command of the *Shackleton*, had become well known to the small community of only 800 residents of Stanley, and was popular. He was tall, good-looking and athletic. He rode a horse well, was seen to ride out of Stanley over the camp, and would show off that prowess as a jockey at the Boxing Day races. He had also formed a relationship with the lady owner of the Rincon Grande sheep station. All of these things endeared him to many islanders; although there was no doubt about his obvious misdemeanour, many ashore without the proper facts, but with unlimited rumours circulating, took his side and thought that we, his subordinates and likely to gain from his demise, saw the events as an opportunity to exploit his plight.

In fact, we wanted desperately to be loyal to our captain, as I believe most merchant seamen are, recognising their senior officers' responsibilities. He had treated us well and we had enjoyed being under his command; so being quizzed and answering questions as to his behaviour aboard was very distasteful for the whole ship's company. We deck officers were very aware that if any of our answers were perceived to give any hint of criticism of him, it would hurt Brown and improve our chance of promotion. Similarly we felt a reluctance to answer some questions, because not one of our interrogators was in any way connected with the sea and knew little of what they were asking, or how to phrase it. We aboard, and many ashore, began to be torn apart as the truth behind the incident was sought. Were his actions the result of drunkenness or temporary insanity? Was our evidence truthful or self-seeking? Tom Flack, the chief officer, was naturally particularly susceptible to outside criticism and was very conscious of, in fact disturbed by, the delicate position he found himself in. A further matter for him privately was that although he was a very capable officer, holding part of his extra master's certificate, and being older and more widely experienced than Brown, and in many ways more mature than him, he was not cut out for a life in polar waters. I do not believe he actually wanted the responsibility of command south. He was thus torn two ways: a very straight, honest man who found it hard to defend his master's ultimate actions but almost certainly, partly for his own reasons, did not want to see him sacked. The enquiry panel seemed obsessed about, and therefore primarily focused on, our behaviour aboard, particularly drinking, no doubt coloured by our visible excess in that direction whilst in port and by the layman's perception of drunken seamen. Nothing could have been further from the truth; at sea we were diligently abstemious. Yet any culture of heavy drinking aboard and the question of sobriety, even on the night of being holed, was repeatedly raised. It was as infuriating as it was unfair and untrue; and the matter of the master's mental health was barely considered. However, regardless of the route the enquiry took, the result was inevitable. Brown was relieved of his command and soon made his way from Stanley to stay with his friend on her sheep station at Rincon Grande. I never saw him again or even heard definitively of his whereabouts for he moved on from the Falklands, some said to South America.

We sailed, yet again, for South Georgia with Tom Flack in command and myself as chief officer after only a few months as second. But the enquiry into the accident and Brown's subsequent demise followed us there, once more in the form of the Colonial Secretary, but this time heading up a new team of inquisitors which included a mariner and a stenographer, all embarked aboard HMS *Protector*. They met us at Grytviken, and the questioning started all over again, mainly because friends and family of Brown had mounted a vigorous defence of him in the UK. It made no difference to the outcome, and if one purpose of the further investigation was to look more closely at us in anticipation of finding a poorly managed and badly behaved ship, they found the opposite. The well-executed workings of our little vessel made themselves apparent, particularly through the viewing of our meticulously kept log books, hydrographic notes, running, sketch and full surveys. The enquiry closed, and with considerable relief but shorthanded by one deck officer, we conducted the final tour of the bases then returned to Stanley. Here unfortunately a nylon mooring line fouled the constantly turning variable pitch propeller when berthing in a gale.

The tail end shaft revolved in oil as a coolant, and the rope had damaged the gland at the after end of the stern tube, causing the oil to be replaced by seawater, which boiled when we ran at full speed. This meant that we had to make our way slowly to Montevideo to have it repaired. The occurrence was probably partly brought about by our being shorthanded and having no deck officer aft at the time. Yet in the eyes of the governor it confirmed his poor regard for our running of the ship as a merchant service entity, and strengthened his resolve for her somehow to be commanded by a naval officer.

During the last call of the season at Port Stanley, when we loaded most of the water-damaged stores that had been sorted there for return to the UK, the governor spoke to Tom Flack; he made it quite clear to Tom that he was only acting captain of *Shackleton*, and that he had other ideas for her command.

Plane down – soundings up

Tidebound alongside the Public Jetty, preparing to sail home, we were advised that one of the Falkland Island Government Air Service aircraft, a single-engine de Havilland Beaver, had flipped during take-off on water around the camp. She was in the small harbour of El Moro, adjacent to the Rincon Grande settlement, 60 miles distant in the East Falklands. We were requested to proceed there to attempt salvage. We arrived to find that she had been moored quite close to the station's small jetty, but was upside down and waterlogged. Even at some distance off to work we were nearly aground at high water, and Tom Flack, now in command and fearful of becoming stranded there, decided upon a speedy recovery. We hauled the aircraft towards us but then, rather too hurriedly, lifted her without allowing time for sufficient water to drain from her wings and fuselage. We got her on deck the right way up, but although she was intact she was so badly strained that despite her return to Stanley in one piece she never flew again.

As we left our anchorage with little way on, we ran into a patch of soft mud in the centre of the small cove. As we were doing little speed it did no damage, and we quickly swung into our surveying mode, re-anchoring to put the launch down. Sounding round, we found there to be a knoll with one fathom of water over it whereas all around, before shoaling towards the shoreline, there were four fathoms. We positioned the knoll on a quick sketch survey to be submitted as a hydrographic note on our return to the UK, along with our other findings of the season. We had used a chart that showed five soundings in the area of the cove in a 'five spots on a dice' configuration, each indicating four fathoms. Like the majority of those of the Falklands at that time, that chart was based on the 1838 survey by Sullivan, who was first lieutenant to Captain Fitzroy aboard HMS *Beagle*. I personally took our survey work from that voyage, including that sketch survey, to the Hydrographic Office in Cricklewood. I was greeted warmly, for we had established an excellent relationship with that office, and was invited to view some of the charts in their archives. I had on previous visits seen and marvelled at the detailed original work of Cook, Dalrymple and other great surveyors. On this occasion a folio was produced showing Sullivan's original survey for Salvador Waters and adjacent inlets. We were amazed to see that he had drawn the cove with five soundings disposed exactly as on the current chart, except that the central sounding bore the figure one in the style of the time with a flat base and sloping projection from the top to the left. Sullivan, in an open boat with a hand lead line, had got it right in 1838! The modern cartographer, compiling the recent chart had lifted that central figure indicating one fathom, as a further four (fathoms), not correctly as the six-foot patch that we had found.

Finally, in Stanley, before our departure home and after a drama-filled voyage, Tom Flack unsurprisingly arranged for himself immediately upon our arrival in the UK, to return to the Falklands to become the relief master of the Falkland Island Company ship *Darwin*. His departure was not entirely unexpected, for he, the academic who had joined the ship in Denmark with Adam and me, had been rebuffed by the governor, was never really suited to the style of life or tasks down south, and whilst he would have been very competent in that role, I was sure that it was not the command he would have chosen. He had been a pleasant and sound companion, and his going left us with another hole, this time in our ranks, and placed the Survey in a predicament.

Tom, though, had not made his final contribution to our affairs; he gave much thought to the question of our steering. He believed that testing alongside or steaming in placid waters was not sufficiently robust. During the homeward passage we conducted trials with the ship's head swinging firmly to port and starboard, then putting the helm in the opposite direction to correct the swing induced when making way at moderate speed. We found that when swinging to starboard the corrective helm to port worked very well, but when swinging to port the response to the corrective helm to starboard was extremely poor. We knew not why, but knew well that it was the port side that had swung into that floe and holed our ship in the Scotia Sea.

6 RRS *Shackleton*: Refit and Fourth Voyage

One problem solved, another posed

On our return to Southampton, a large conference chaired by a senior officer of the Colonial Service was convened to review our accident and to make proposals to improve our stability and integrity. Mr Cox from the Ministry of Transport had the results of the enquiries; he stated that the master and crew could not be held to blame, and declared himself to be in agreement with the majority of Tom Flack's calculations as to our precarious position in the ship's flooded condition. Uncomfortable exchanges were had regarding responsibility for that, but consensus was reached on the structural changes required, as detailed earlier, as was the upgrading of our lifesaving appliances from that of a cargo vessel to that of a passenger ship. An engineer superintendent from Royal Mail Lines engaged to look into the performance of the steering gear declared it to be in good working order. This I disputed as his tests had been made alongside the berth, and referring to Tom Flack's trials at sea I urged a further investigation. Some time after the conference we underwent more trials, and the same superintendent reported to the shipyard, as if it were some minor matter for rectification, that it had been found that a relief valve on one side of the system was operating at a considerably lower pressure than it should have been, thereby prohibiting a correction to a swing when required. Finally the matter had been solved and Brown exonerated of having committed even a slight error of judgement, yet few, let alone he, were made aware of it.

Because I had not yet gained my master's certificate I could not be given command. But even had I managed to sit and pass that exam, at 25 and with only three seasons' Antarctic experience, and of that only months as chief officer, it was unlikely that I would be considered for the post. Johnston being the only holder of a master's certificate in the organisation and in command of the new *John Biscoe*, the dilemma that therefore arose for the Survey was that no British master mariner with polar experience, let alone one who had sailed as master, or was suitable to take command of *Shackleton*, could be found. I worked in Thornycroft's yard almost throughout the summer of 1958 as the only deck representative from the ship, until latterly a second officer arrived to assist me.

However, the firm of naval architects that had been responsible for the original modifications of the ship in Denmark, now run by younger and more up-to-date partners, particularly the likeable Bob Cree, oversaw the refit in conjunction with the Crown Agents, who monitored the expenditure. I also paid visits to the Admiralty Compass Division at Slough to discuss the abysmal performance of our Sperry gyro compass. During the first voyages two models had frequently failed, sometimes for extensive periods. It was high latitudes, quick changes in temperature and shock that threw them. But high latitudes were our working environment; the gyro compasses being sited amidships in the wheelhouse temperature changes came from the necessity to open the bridge wing doors; and shock resulted from working ice. One, or any combination, of these three resulted in large precessing[33] errors or total failure. Difficulty in maintaining a course without continually resorting to the magnetic compass – which in our high southern latitudes had large errors and poor directional stability – was almost the norm. When surveying, turning at the end of lines, and particularly when standing into the coast at right angles to the contours then picking up a reciprocal parallel course a short distance to the side of the one just completed was a nightmare, and had frequently resulted in the ship skewing off at an angle into danger. Slough came up with a possible answer, an Arma Brown model designed for use in tanks. Such a compass was installed, but not being designed for the larger environment of a ship, it could not transmit to the slave repeaters used for taking bearings on bridge wings.

The new second officer, Jim Martin, a Scot from Jedburgh, began to make his mark as an enthusiastic and likeable colleague, and at a late stage a third officer joined whilst the Survey still continued to search for a master. Amongst their many varied capacities representing the Overseas Territories, the Crown Agents were then seeking a whaling inspector to be based at South Georgia. A retired naval captain in his fifties applied for the position and knowing the plight of the Survey, they sent him for interview. On paper he was impressive, having distinguished himself during the war and been awarded the DSO and Bar, and at interview, I was told, appeared a good candidate. The governor vociferously encouraged such an appointment as it conformed entirely to his idea of how we should be commanded. By this time the Survey would have been desperate and after the debacle with young Norman Brown, the idea of an older and hopefully more stable man, even though without polar experience, would have appealed. He was expecting to take his wife to South Georgia, that being within the terms and conditions of agreement for that position, so in offering him the position of master of the *Shackleton*, the Survey also made the offer for him to carry his wife aboard as far as the Falklands.

He accepted the position and I met our new master when he first visited the ship in Southampton, about ten days prior to our departure. Jim Martin and I were preparing the ship for the forthcoming voyage, and I happened to be on deck when he came up the gangway. He asked me if I smoked cigars, to which I replied, 'Yes, I like small cheroots,' expecting to be offered one, to which he replied 'So do I; I will have some put in my bond.' That was his entire contribution to our preparations. He was a big man, well over six feet, distinguished looking, of upright bearing, but very stiff of gait. He had an educated voice and was affable, but changed distinctly to not being so communicative when his wife was nearby.

---

33 The compass rotating from its correct alignment.

Fourth voyage: another unfortunate occurrence

We sailed from Southampton for the ship's fourth season south, and on clearing the Needles, Jim as navigator gave our new master the course down the Channel to pass off Ushant. He repeated it to the helmsman and went below. No queries as to the compass error or about the last-minute stowage of the deck cargo. No words of advice, let alone instructions regarding traffic, and no night orders. Nothing.

We began to wonder, but we behaved impeccably at first, calling him for landfalls and major alterations of course, telling him it was normal to make weekly inspections of his vessel under the Merchant Shipping Acts, and informing him of the requirement for boat and fire drills. He occasionally took an interest, and we thought 'this is a very straightforward passage down through the tropics to Montevideo – he must regard it as mundane after his war service and will spring to life when necessary'. It was, and he did, for we unexpectedly had reason to call at the Cape Verde Islands.

Grand Harbour on São Vicente is a large bay where the main port was situated, which was once busy as a coaling station. The islands are volcanic and steep to. I had told the captain of our landfall, and he came to the bridge as we entered the bay. I suppose deep down our suspicions were much greater than we cared to admit, for it was interesting that we had already departed from the then time-honoured merchant service form of mate on the forecastle, second mate aft, and third mate assisting the master on the bridge or about the deck preparing gangways and boats. I, as chief officer, was on the bridge, and Jim was on the forecastle with the anchor party. Suddenly at little less than full speed, for we had only just put the engine room telegraph to Stand By, and in over 100 fathoms, the captain took the loudhailer to the bridge wing and shouted, 'Let go the anchor!' Unfortunately the bosun's mate was at that very moment preparing the anchor for dropping and had his hands on the windlass brake; before Jim could intervene the anchor was dropped and, sparks flying, the cable streamed out madly astern. Fortunately at that depth the anchor never touched the bottom; had it done so we would have lost it with the cable torn violently from the fastening in the cable locker. I thundered astern, as Jim and the bosun's mate miraculously managed to arrest the cable, but the windlass was then unable to retrieve the full weight of cable and anchor. Eventually we managed to do this by repeatedly steaming into shallow water, hauling the cable with us, so that the anchor and much of the cable grounded, thus relieving some of the weight, enabling the windlass to lift it gradually back aboard. The port authorities awaiting our arrival were very suspicious of our clawing around the shores of their harbour, and once we had berthed became particularly unpleasant.

As we sailed southward again we realised that our captain was not just going to be a poor master, but was going to be a downright liability and would require watching at all times. He and his wife drank continuously. He appeared to carry it well when on view, speaking slowly but clearly, but at times we would not see him for days. His wife was worse, but we saw even less of her except when she caught us! Occasionally I went to their cabin, for I still believed that as the master he should know something of what was going on. The other officers got drawn in when passing the door to the captain's quarters, sited as they were at the foot of the companionway to the radio office and wheelhouse. When we were called in, a drink would be thrust in our hands, and then, on wishing to leave, thinking we had done our duty and humoured them, she would pour more, usually gin, through our fingers as we covered our glass and said 'No thanks.'

Shortly after leaving the Cape Verde Islands nearly all aboard were struck down by the Asian flu which had been prevalent in the UK at the time of our departure. We became so shorthanded in all departments that we were forced to anchor off Dakar until sufficient numbers had recuperated to run the ship. There were some notable exceptions to those who succumbed to the virus, the captain and his wife being two, but more importantly in the context of keeping a modicum of services running was the donkeyman, the leading hand in the engine room. Staying healthy during the height of our infection, he not only kept the auxiliary machinery running, but took over the galley, from which he fed us single-handedly.

On our arrival in Stanley the captain came to the bridge, stiffly erect in uniform with cap, and took over. Unfortunately, at this stage I was still not bold enough to deny him the handling of his own ship. As we closed the jetty he began to give orders that were either meaningless or unacceptable, and by the time I did step in by telling the helmsman to listen to me only and by handling the variable pitch propeller control lever quietly myself regardless of the captain's rantings, it was too late; the die was cast and the inevitable loomed. We overran the corner of the small jetty and brought up aground alongside the town's sewer outfall pipe that extended out from the shore just beyond. Fortunately we had managed to drop the anchor in roughly the correct distance off the berth, and recovered the situation. Survey colleagues and the harbour master/customs officer came aboard and went up to the captain's cabin. As the master dispensed drinks and his dishevelled wife fell about, the chief steward and I sorted the official business, whilst no one else batted an eyelid as to what was so blatantly amiss.

That first arrival in our homeport was also the last time that the master handled the vessel at a berth, as a seven-month charade then commenced. Often he came to the bridge as we prepared to dock; sometimes he issued orders that we ignored, and sometimes an individual was detailed off to engage him in conversation so as to distract him from being inclined to interfere. As I write those last words, I still feel uneasy and find totally abhorrent the actions we were forced to take. The captain and his wife found accommodation ashore in Stanley, and when there we saw little of him aboard, although he pulled himself together sufficiently to pay a call on the governor. She soon gained a reputation in the small town, and was frequently to be seen trundling her shopping trolley to and from the Company's West Store with bottles.

We sailed south with the effigy of a captain upright on the shoreward wing of the bridge, but without his disruptive wife who remained ashore. The ship settled into a highly unusual but satisfactory mode of operation. Jim Martin, the lanky Scot with a great sense of humour, always cheerful and willing, was a tremendous asset to the ship. The wrong person as second officer could have exploited the situation, made it very difficult for me or caused dissent throughout the ship. As it was, he was a tower of supportive strength, assuming a role far beyond his years in keeping both officers and crew loyal to me. We got on extremely well together, sharing many of the responsibilities and decisions, and indeed much of the humour that arose from the situation, with me practically in command and conducting the programme, handling the ship and the ship's business. We were only 25 years old, both holding just our first mate's certificates. We carried only one other deck officer, who held a second mate's certificate, and between us we had to carry the entire responsibility for every aspect of the ship. I also had to liaise with the Stanley office and bases over the radio as to our programme, work out the crew's wages, and order fuel and stores, all in the master's name. It was demanding, but matters weigh lightly on the shoulders of the young.

Two things made it all not only possible, but a real success. Firstly, which I was aware of at the time, the unflagging loyalty and support of Jim Martin and the entire ship's company. Secondly, which only in hindsight occurred to me, for I never gave it a moment's thought at the time, that the ultimate responsibility was never actually mine. There was a captain aboard, and should things go awry it was he who carried that responsibility. It was a tremendous voyage for me; I gained invaluable knowledge and experience that otherwise I would never have acquired until many years later. I was bold, but I was lucky. There must have been many near misses that I was unaware of, but some of my mistakes during that voyage still find their way into my dreams even now. I ran wide of a narrow channel manoeuvring at too great a speed in fog amongst some islands off South Georgia. Only when becoming surrounded by kelp did I realise how close to the shore we were, so I stopped, took stock and proceeded much more cautiously. Similarly entering Potter's Cove in the dark, a comfortable anchorage but with a narrow approach, I was fooled as to the line of the channel by grounded ice on one shoreline. I found the ship so close to the opposite shore that it took my breath away. Impatiently going astern in ice I bent one of the propeller blades, but this was not too uncommon an occurrence. And so the voyage progressed, but fortunately with rather more acceptable acts of seamanship and behaviour by myself than the events related above.

The master did have some good moments of clarity, but was hopelessly out of his depth in all that was going on. He came to the bridge as we left Admiralty Bay, a wide deep straight inlet culminating at its head in a cloverleaf configuration of bays, each having its own glacier and spectacular in its own right, but similar. He was itching to take over. There was no sea ice and it was lunch time, so providing him with the courses to be followed and setting him off in the right direction we left him alone, like any other watch-keeping officer. But then, on frequently getting up from the wardroom table to peer forward through the ports, we soon saw that he had taken a wrong turning and was sailing into one of the other head bays towards the glacier rather than down the main bay and out to sea. We had expected it, and it was perhaps both cruel and puerile to allow him to display his incompetence in such a manner, but it was also somewhat of a test of his ability and thus a reassurance to ourselves that we were acting correctly in our general treatment of him. Also, occasionally having to do his job as well as our own, slight resentment would bubble to the surface in the form of a prank. This was not the case in the next incident, which had a major effect on the remainder of the voyage.

In mid-December there was an abundance of heavy pack in the Bransfield Strait, which Jim and I had worked throughout one night and into the following morning. We were on our way to Nelson and Livingstone Islands to support some survey parties, when at 8 a.m., with a long day's work ahead of us, the master came to the bridge desperate to handle the ship, and I let him, for it was convenient to do so; Jim was asleep and I needed my breakfast. There could hardly be a seaman who would not want to try his hand at twisting, turning, pushing and shoving, backing and filling, in the exciting and rewarding manner that is required to progress through pack ice. The captain was no exception; and in tight ice, as it was, there was little likelihood of too heavy an impact, the modus operandi being mostly that of forcing steadily, unless one goes well astern for the hull to create a partly clear run in open water in which to go ahead again to ram.

I handed over to him; what young man would not have given way to his captain twice his age and who appeared to be in control of his senses? I should have said 'No', silently adding, 'Enjoy the spectacle of a calm sunny day in heavy ice with the not-too-distant shimmering mountains and an abundance of wildlife – but leave the ship handling to me' – but I did not and I went below for breakfast. After all, he had seen the ship being worked in ice a good few times at this stage. All went well for ten minutes until, impatient at the poor progress, he backed up astern to have a run at a floe. Made aware of this by the long stern movement of the engines, I went to the bridge believing we might have been stuck at the bow and were attempting to release ourselves, only to find the vessel hurtling stern first towards a heavy floe. The inevitable collision followed. I took over, and on getting under way again it was immediately apparent that our steering was severely defective. We tested as well as we could, having found some relatively open water, and by trial and error found the rudder to be lying well off centre with the wheel amidships. Pumping out the after peak[34] and ballasting forward, we brought the top of the rudder easily within sight and confirmed it to be bent by about 25 degrees. Either the rudder stock or the blade was twisted. The engineers set to in the steering flat[35] to offset the quadrant at the rudder head and realign the chain tackle that was driven by the hydraulic system emanating from the wheelhouse. However, they could not alter it sufficiently to give us enough degree of helm both sides of amidships to complete the voyage without endangering the ship. My thoughts turned to ramming the rudder against the ice again to bend it back into position. Although that had sometimes successfully been carried out, Captain Maro taking the *Theron* into the Weddell Sea for the Trans-Antarctic Expedition advance party having managed to do so, I was none too keen to do that, fearing weakening or, worse, fracturing the twisted metal of the rudder stock, which was where the damage appeared to be. Also the angle at which we could position the rudder for it to be bent back in place by going astern onto ice was too far away from central for me to be confident of not bending it further the wrong way. Nor did the idea of a re-twisted rudder stock to complete the season endear itself to me, and if I did not want to attempt to finish the season in such a state, why risk it at all?

---

34 The ballast tank contained within the stern.

35 Compartment housing the steering gear.

So, with the quadrant and rudder set over as much as possible to make steering possible, although difficult, we made our way northward towards Montevideo where the nearest dry dock lay to accept us for repairs. The 1,800-mile passage, with a pause at the Falklands which lay en route to disembark some stores and personnel, was completed quite easily but not very efficiently, our wake looking as if we had a drunken helmsman as we yawed from side to side. Montevideo is approached by way of a five-mile dredged channel through the muddy bottom of the River Plate, with the masts of the *Graf Spee* lying to port identifying her last resting place after Captain Langsdorff had scuttled her rather than face the *Ajax* and the *Achilles*. When a vessel is close to the bottom steering is usually adversely affected, and our damaged rudder made it even worse, especially noticeable amongst outward shipping, but with the appropriate signal flying, and latterly the assistance of a harbour tug we berthed. We waited nearly a week for the dry dock to become available, but once we were in, it was confirmed that the rudder stock was twisted by 28 degrees. After consultation with the shipyard and Lloyd's, the London office decided that a new stock should be forged in the UK and flown out to us in Montevideo. The yard at which the ship had been built in Sweden provided the requisite drawings, which at some stage were copied before a set was sent to the chosen manufacturer in the UK to forge a new stock for us. It took about another three weeks after we had entered dock for the large piece of steelwork to arrive. It was 18 inches too short. The explanation was that the drawing, when copied, had been folded under at a break in the lines of the parallel sides of the stock, and that all beyond the fold had not been taken into account. The ship's company had already settled into a life of pleasant integration with both the very pro-British Uruguayans and the ex-pat community. The captain and his wife found accommodation ashore, understandably, for living aboard in dry dock, although all services were maintained, was not ideal, and in truth most of the initial period there was spent just waiting.

Midsummer in Montevideo could hardly have been more pleasant. The weather was almost perfect, just two or three times being interrupted by a most interesting phenomenon, a *pampero*, which I had never witnessed because we did not usually visit in midsummer, when they occur. A *pampero* is a severe line squall travelling just ahead of a cold front associated with a depression, travelling west to east over the southern Andes. Losing much of its moisture as it does so, the now cold dry air, borne on a strong southerly wind, meets and cuts under the warm humid air further north. The clash produces wind, rain, thunder and lightning – and unusually and slightly alarmingly, it also brings with it millions of all manner of flying insects, ranging from large locust-like creatures to moths of all sizes and minute bugs. These are all lifted from the plains to the south of the River Plate and driven ahead of the front as it progresses northward. They fill the air, and are particularly noticeable at night, swarming around any light.

The ship's providor and his wife frequently took us to their beach-side house, and they and the ship's agents also took us to the Anglo-Uruguayan Montevideo Cricket Club, which was essentially a country club with tennis, swimming, croquet, bowls and *pelota* facilities, and a vigorous social scene. The cricketers amongst us, practising in the nets, managed to get a game when required in one of the MCC's teams, and it was soon realised that we had the makings of a good side ourselves, our bosun's mate being an exceptionally fine fast bowler. We eventually played against their first eleven, and won. The local paper reported it and this was picked up by the English edition of the *Buenos Aires Herald*, which in turn was picked up by the *Daily Telegraph* reporting that RRS *Shackleton* had beaten the MCC in Montevideo! An MCC touring side had actually been in South America about that time. My father read the report and wrote out to me: 'Well done son, I did not think you were that good.'

The second rudder stock in the correct proportions arrived and was fitted, and at last after many weeks we were set to go. As was the custom in South America, for any major expenditure on behalf of the ship, the captain, and in this case the chief engineer, the ship being in a repair yard, received a substantial handout from the manager of the yard. The captain, who had only returned to the ship the previous day and hardly knew the manager, was completely taken aback by this. He asked me what to do with the money and I was taken aback by his innocence. I told him not to return it, for it was always factored into the price of the repairs and would only go into someone else's pocket. 'Put it in the various mess funds,' I said – or better still, as everyone had spent greatly during the preceding weeks in port instead of saving their money whilst working south – 'Split it amongst the crew'. He did both, and they all thought what a good chap he was.

Well past the midpoint of the season we sailed south again under the command of our captain, as I wondered whether or not this extended time in port, guaranteed by the lengthy procurement and fitting of a new rudder stock, might have been the ideal opportunity to replace him. Having seen him drunk and incapable on a number of occasions, did those in authority in Stanley think that he was safe when south? Did they honestly believe that he could be so drunk in port yet sober and capable the moment the lines were let go? The ship came and went, the programme was executed, and so the situation was, disgracefully, allowed to continue, presumably in the hope that all ultimately would be resolved when the voyage finished. Did the Stanley authorities never even advise the head office of their apprehension, fearing another debacle involving a disgraced captain were this one to be relieved of his command, the second FIDS master to be so in as many years? Was the director or our London office ever made aware of the problem? It would appear that the lack of action was generated by a need to save face. What an extraordinary gamble with the lives of young men! In Stanley there was a culture of drinking, both with the islanders and the ex-patriots, and by many an acceptance of it that extended to visiting ships, whose crews were by no means abstemious when in port. There was also perhaps a familiarity with tales of drama from the Antarctic, Stanley having always been the rear base for the Survey and on the route of many expeditions, and again there was an acceptance of individualistic and eccentric behaviour by polar adventurers.

But there were those there who should have been above such thinking. To this day I do not know the answers to all the questions, but do believe that the governor at the time, who had witnessed us nearly sinking on his first foray south and knew of the dangers of working there, must bear the greater proportion of the blame for the turning of blind eyes. My own behaviour can be deemed not entirely satisfactory; I should in hindsight have spoken up, but whistleblowing was not the vogue in those days whereas loyalty to one's superiors was – and especially so at sea, when many very worn-out merchant service captains who had survived a terrible war were still in command, maintaining their position entirely through the support of their juniors. I knew too that I was benefiting enormously from the episode, for it changed me from a young novice sailor to a reasonably competent mariner. It was, however, a disgraceful period in the maritime history of FIDS. Fortunately there was to be an event later that for those on board gave some satisfactory closure to the sorry saga of our captain himself, when set apart from the extraordinary events of the voyage.

Tristan da Cunha, St Helena, and closure to an episode

We called at Stanley before returning to the Antarctic Peninsula to complete what was left of the season south before commencing the passage home, as usual passing through Stanley and Montevideo. At the latter we were co-opted by the Colonial Office for what seemed to appear an interesting variation to our normal run home. It proved to be all of that. A Bank Line vessel had transported a complete sewage system to the island of Tristan da Cunha in the South Atlantic for instalment at its only settlement of Edinburgh, but she had lain offshore for some 30 days unable to discharge her cargo, mainly on account of the swell. Her charter period having expired, she then made for Montevideo, where she left the mass of pipes and equipment on the quayside. We were asked to load them and proceed home via Tristan in the hope that we would be more successful in landing them.

The voyage of about 1,000 miles almost due east was of little consequence other than that the captain's wife, having re-embarked in Stanley, drank herself into what I would describe as paranoid delirium. She rampaged around the alleyways of the officers' accommodation and the adjacent open decks in what appeared to be a pink nightdress, claiming very loudly that we had set her husband adrift in a small boat, but that unlike Captain Bligh he was supposedly alone. The incident might sound amusing now, but in fact was most alarming for we feared that she could fall overboard at any time, trying to espy her husband. By the time the near-7,000-foot volcanic peak of Tristan hove into sight above the clouds, many hours before we saw the base of the island, our doctor had fortunately sedated her, and we were able to focus on the island, its extraordinary inhabitants and the task in hand.

The Tristan de Cunha Island group is part of the Mid-Atlantic Ridge; the largest island bears the same name, the others being Gough, Nightingale and Inaccessible Islands. Tristan da Cunha Island, only six miles in diameter, is a volcanic shield cone of 6,760 feet in height, with Mary's Peak rising from a surrounding high plateau scored by ravines down which torrents form in heavy rain. This plateau in turn drops abruptly into the sea, with basalt cliffs of up to 2,000 feet in elevation. Rarely is there a break in the cliffs, but at one point on the northern shore line there is a break, forming two beaches; adjacent to these and behind lower cliffs, is a narrow shelf of grassland barely a mile in length and half a mile in width. Here the settlement is located.

Discovered and named after himself by a Portuguese explorer, Tristão d'Acunha, in 1506, who, like many visiting after him was unable to land, the island was eventually inhabited by an American, Jonathan Lambert, between 1810 and 1812. He drowned there, and it was not occupied again until Britain annexed it in 1816 by placing a garrison of marines on what is claimed to be the remotest island on the planet. The marines were probably put there with the intention of preventing the French from utilising it as a base from which to mount a rescue of Napoleon, who had been imprisoned on St Helena some 1,500 miles to the north; the marines were withdrawn soon after his death. One of them, however, William Glass, returned with his wife and children and founded the colony whose descendants were to meet us. More settlers arrived before the turn of the century, notably Thomas Swain, the seaman who had caught Nelson as he fell at Trafalgar, whilst survivors from shipwrecks added new blood to the small contingent. Peter Green,[36] a Dutchman, from the wreck of the schooner *Emily* in 1836 remained there, falling for and marrying a local girl, anglicising his name, and creating a mini-dynasty. Augustus Earle, who later came to prominence as Darwin's artist aboard the *Beagle*, came ashore for one day's painting in 1824 from another vessel, and unable to re-embark due to bad weather, stayed for eight months. His progeny, if any, are not to my knowledge recorded. In 1827 HMS *Duke of Gloucester* on a visit was asked to return with ladies from St Helena to be brides for the spare bachelors. This she did and the plan was apparently successful; when we arrived, there were only eight surnames amongst the near-200 inhabitants, many of whom we met, all descendants from those early days.

On arrival we were immediately boarded by the head man, William Reppetto, accompanied by several of the elders. Reppetto was a descendant of a survivor of the barque *Italia* that self-combusted in 1892 when carrying coal from Scotland to Cape Town, and was beached on the island. All hands were saved and he, together with a companion, Lavarello, chose to stay, providing two of those eight family names. We quickly realised that the majority of our visitors were aboard for the sole purpose of seeing what they could 'lift'. Not 'steal', for they had always preyed off passing ships, and thought it their god-given right to be given, or if not then take, almost anything they set their eyes upon. We had to watch our stores and anything loose like hawks. We actually caught them taking, but managed to retrieve, a lifebelt, but as it was we were played upon to relinquish some paint, canvas, small cordage and a well-used mooring rope. The discharge of the cargo went well, for the sea conditions prevailing whilst unloading were not dissimilar to working the open coast of the Antarctic Peninsula when ice-free. Handled with care, our motorboat and scow were equal to the swell and used in parallel to the island's longboats.

During the few days it took to land everything we managed to get ashore to the Settlement. The small habitation had been visited in 1867 by Albert, Duke of Edinburgh, in HMS *Galatea*, and formally, delightfully, but somewhat inappropriately, it was named Edinburgh of the Seven Seas in his honour. But Settlement was a much more suitable name for the few dozen houses built from blocks of volcanic tuff and thatched with reeds. Some were built half into the ground, almost windowless, blackened within from wood smoke and smelling of fish, giving a medieval air to the place. The womenfolk sat at their doors spinning yarn or knitting coarse woollen wear, whilst those unoccupied peeked shyly from within or around corners at

36 Pieter Willem Groen

the visiting strangers, scuttling out of sight when conscious of being seen. Potatoes and apples were grown, sheep reared for both their carcasses and their wool, and wood was collected from the mountainside, the men having to climb higher and higher over the years to obtain it, having denuded the lower slopes of trees in little more than a century. The men worked mostly in the crayfish canning factory that had been set up adjacent to one of the bays by a South African corporation. All spoke in a heavy patois and were quite hard to understand; interbreeding was very evident in both looks and manner. A visiting parson was a useful interpreter and go-between. When it came time to leave I was gifted a pair of homespun woollen seaboot socks by the head man as he entrusted a dozen or so of his folk, mostly young, very shy and timorous, into my care for the voyage to Jamestown on St Helena. We had been asked to carry them to facilitate visits to their relatives, links having been maintained with their northern island neighbours since that first occupation by marines from that garrison. We willingly did so, for it provided us with a reason to call into St Helena, almost on the track of our homeward-bound passage.

Our arrival off Jamestown coincided with a visit by the cruiser HMS *Newfoundland* on her way home from the Far East, and we anchored close by her. The navy threw a cocktail party for the local dignitaries, and generously, all our wardroom were invited. It was an altogether more splendid affair than those we were used to aboard HMS *Protector* in Port Stanley; the *Newfoundland* had larger numbers and more space, enhanced by tropical airs beneath smart awnings on her gundeck. Central to this there was a capstan, which also became central to an awful episode. With the party in full swing, our master, by that time well inebriated, took it upon himself to climb upon that capstan, which had been transformed into a bar-cum-servery and, putting his feet into several dishes of canapés and knocking over jugs of gin and tonic, began to make a speech. Removed fairly quickly and firmly and placed aboard our own boat, he was taken back to his ship with minimum fuss but great embarrassment on our part. As the party ended and I was taking my leave of *Newfoundland's* commanding officer, Captain Arthur Hezlet, he surprised me by asking that I remain behind and join a small number of guests he had asked to dine with him. As a 25-year-old chief officer of a merchant vessel, albeit a Royal Research Ship, I could only surmise that the invitation was made out of sympathy for my having had to support an alcoholic captain for so long. I remember little of the other guests, the governor and other local grandees, and the ship's most senior officers, but was enchanted by and well remember my host. He appeared a most knowledgeable, interested, kindly and compassionate man. We talked of exploration worldwide, Fitzroy and the *Beagle,* and touched upon the strengths and frailties of man and his mind, especially during long absences from home and normality, and under stress. He was to be promoted to flag rank on returning to the UK, and eventually became Vice-Admiral Sir Arthur. I later learnt of his heroic exploits as a submariner, in fact one of only a few to have spent the whole of World War II in submarines. In a classic action he had torpedoed the Japanese heavy cruiser *Ashigara* off Sumatra, in the Mediterranean sunk an Italian troop ship, and in the Arctic escorted convoy PQ16 to Murmansk. He had also towed the midget submarine X5 towards Altenfjord for the action against the German pocket battleship *Tirpitz.*

It was true that I had been asked to dinner to relate my adventures of the polar seas, but also quietly to him to talk of my captain, for Hezlet in turn had something to tell me of him. Both had won a DSO and Bar, and DSC amongst other awards in courageous actions. The exact details I forget, but the stories recounted, emphasised by the teller's own record, were of how our captain had several times distinguished himself – outstandingly so, in fact, during the landings of the North African campaign, when more than once he had steamed close inshore to support the landings in the face of heavy fire from ashore. As it was told to me, his bridge and upper works were shot away, but he still commanded his vessel from the stripped-open main deck, raked by small gunfire, whilst the opposing heavy armament was unable at close range to lower its fire enough to direct it at his hull. Regardless of whether the telling of his exploits was exact, or my memory of them correct, it was illuminating for me to hear them from so distinguished a seaman and war hero himself. Our past several months of nursing an often not sober captain were put into context. Appropriate closure was given to the circumstances of an extraordinary voyage; the poor man had been mentally shot to pieces, though apparently bodily spared, and had found his solace in drink, and sadly we had witnessed his inglorious finale. On arrival in Southampton he went down the gangway and I neither saw nor heard of him again, but at least by then I had a better understanding of a once courageous naval officer.

[Above] Coastal scenery. Mt. William, Anvers Island from Neumeyer Channel, near Port Lockroy. (Crown copyright; photograph by John Smith, 1957)

[Below] Surveying from above Danco Island Base, with view of mainland up which no practical route was found to the peninsula plateau. (Crown copyright; photograph by Fred Wooden, 1956)

Chris Elliott (later captain) taking motor boat and scow ashore at Adelaide Island laden with Muskeg tractor. (NERC Copyright; photograph by BAS Film, 1971)

Chris Elliott embarking huskies in an inflatable craft at Adelaide Island for transfer to Stonington Island. (NERC copyright; photograph by Dave Rowley, 1972)

Hope Bay showing ice-foot utilised for landing, and Mts. Flora and Taylor between which excessive Fohn winds can blow. (Crown copyright; photograph by Frank Preston, 1962)

John Biscoe approaching Argentine Islands Base through Meek Channel with two American icebreakers lying in Penola Strait. *Biscoes'* track through ice shows route close to ice cliff necessary to avoid rock in centre of entrance. Mainland Peninsular across Strait five miles distant, where pyramidal Cape Tuxen is 2000 feet high and mountains beyond rise to about 8000 feet.
(US Navy copyright, 1959)

Tein peaks (2500 ft in height) of Cape Renard at entrance to the Lemaire Channel.
(Crown copyright; photograph by George Larmour, 1958)

Stores from ship being moved to base, Hope Bay.
(Crown copyright; photograph by John Thorne, 1957)

Deck Officer taking bearings from *John Bisoce* in Neumeyer Channel.
(Crown copyright; photograph by Tom Woodfield)

[Above] Huskies preparing to haul sledges from *John Biscoe* beset off Adelaide Island. (Crown copyright; photograph by Tom Woodfield)

[Below] *Bransfield* lying on edge of wind driven shore lead off Adelaide Island, and author taking stroll on fine day. (NERC copyright; photograph by Ella Woodfield)

[Above] *John Biscoe* beset amidst large floes in the Penola Strait. An opportunity to clean up her name. (Crown copyright; photograph by Colin Johnson, 1958)

[Below] *Bransfield* entering Borge Bay anchorage, Signy Island. Note Small Rock close to starboard overrun by light ice, Bare Rock to port just showing above ice, both with men adjacent. Coronation Island in distance. (NERC copyright; photographer unknown, gifted to author)

Peninsula coastal scenery (NERC copyright; photograph by Douglas Brown, 1967)

Stonington Island Base. (Crown copyright; photograph by Tom Woodfield)

Borge Bay, Signy Island showing rocks through which Biscoe had to pass during night of storm (Crown copyright; photograph by Gordon Robin, 1947)

[Above] Landing field party utilising their own dogs to transport stores and equipment onto the Jones Ice Shelf, northern Marguerite Bay. (NERC copyright; photograph by Tom Woodfield)

[Below] *John Biscoe* feeling dwarfed passing between two icebergs whilst surveying in the Woodfield Channel (Crown copyright; John Killingbeck, 1963)

7 RRS *John Biscoe*: My Voyages as Chief Officer

In the summer of 1959 I was still not qualified to take command. However David Turnbull, a New Zealander with an extra master's certificate, who was an examiner of masters and mates, and a surveyor of shipping with the Board of Trade in London, had been recruited a year earlier to sail as chief officer of the *John Biscoe*. He was now appointed master of the *Shackleton*, and I, having first managed to obtain my master's certificate ticket, took his place aboard *Biscoe* under Captain Johnston. The latter tended to look after his own interests, and rather than letting the Survey keep me aboard *Shackleton* to assist the fairly new Turnbull, he had asked me if I wished to transfer to his ship, the three-year-old *John Biscoe*, to strengthen his team and take some of the load off his shoulders.

Johnston, as I knew from my first voyage south on *Shackleton* four years previously, and innumerable reports from others aboard *Biscoe* since then, was mostly aloof, a hard taskmaster, difficult to please, and uncompromising in the standards he set and required aboard. Yet he was exciting to sail under, and from him an enormous amount could be learnt. He was the only ship master ever to have spent several continuous whole seasons in Antarctic waters. He was an exceptional seaman, and although extremely adventurous in the pursuit of the ambitions and goals of the Survey and the individual young scientists aboard under his control, his judgement was such that he always remained fractionally on the right side of the line between success and disaster. For him, disaster south would have meant the loss of the ship or rendering it totally incapacitated, whereas damage from ice, temporary stranding in hurricane-force winds and gently clipping rocks were almost routine, and certainly part of the job as far as he was concerned. Later, in a more sophisticated vessel, I came to modify that view for myself, but he was to teach me patience and coolness in difficult circumstances, and he helped me develop my judgement through observation of his practices. I knew well of his style, leaving the deck officers to carry out the majority of bridge duties such as handling the ship in ice, only to take over when extreme conditions dictated he did so. I therefore jumped at the chance of serving under him again, although I well knew that the freedom of the major decision making and the near command I had enjoyed on the previous voyage would be gone, to be replaced by being within his dictatorial regime. A further incentive for me to move across to the *John Biscoe* was that she played the major, more demanding, southerly role of our two ships in the more severe ice of Marguerite Bay and the Weddell Sea, whilst the *Shackleton* had begun to undertake a great deal of geophysical and oceanographic work in the open seas north of the peninsula.

True, also, was the fact that David Turnbull had already, in one year as chief officer of the *Biscoe*, gained the reputation of being a difficult person with whom to get along, so sailing under him aboard *Shackleton* was not a welcome prospect. David, aged 38, from a family of shipping agents, farmers and wool merchants in Timaru in New Zealand had, like me but much earlier, served his apprenticeship with the Port Line. After the war in various convoys, he had joined the Union Steam Ship Company of New Zealand, then Holm & Company, where he experienced farm-related seafaring around the Chatham Islands, similar to that experienced around the Falklands. He got his first command at age 32, and gained his 'extras' at 35. He was undoubtedly a good practical seaman with a sharp intellect but with a very disconcerting, sometimes awkward and unhelpful manner, and a sarcastic wit; he could be generally disagreeable. He had soon become nicknamed Frosty, yet I got on with him well – albeit distantly at first, we being of different rank on different vessels – but later, as my fellow master aboard our two small ships south, very agreeably. He remained master of the *Shackleton* until he retired from the Survey in 1970.

The *John Biscoe*, built by Fleming & Ferguson on the Clyde, and launched in 1956, was 220 feet in length and 40 feet in beam, and had a loaded draught of around 18 feet. Of 1,584 tons gross, she was of diesel-electric propulsion delivering 1,400 horsepower, giving a service speed of 12 knots, with a range of 18,000 miles. She had a single hull with double bottom, and wing tanks aft for fuel. Built beyond Lloyd's Ice Class I Classification, of half to three-quarter inch, low-temperature, high-impact steel plating, she initially had single frame spacing of 24 inches. After severe damage on her first voyage, she was strengthened with intermediate frames and deep longitudinal stringers.[37] She was fitted with two 10-ton safe working load derricks for lifting craft and working cargo. She accommodated 32 crew and 35 Fids.

The requirement for an ice-strengthened cargo-carrying vessel with similar service parameters to those required for the *Shackleton* was a task none too easy to achieve, but one tackled well by Graham & Woolnough, the naval architects that had overseen her modifications; and a capable vessel was built. There was little if any polar expertise to draw upon in Britain, but there was abundant knowledge abroad on topics such as: Mayer-form icebreaking bows; the protection of stern gear from ice; the maintenance of integrity when damaged; the ability to roll to break the friction between hull and ice by the provision of pumping across wing tanks; the best delivery of power to achieve icebreaking speed from a standing start when backed up in ice; and the equipping of a proper crow's nest from which to handle the ship in ice. All such matters were available to be researched and included in *Biscoe's* design, but few were. How much better this good little ship could have been had it not been for one obstacle. That was the ultra-conservatism of Bill Johnston, yet it was he who knew better than any the conditions to be dealt with down south. He vetoed the incorporation of every modern development; did not believe in the practice of rising up on pack ice by means of a sloping bow to break the ice with the weight of the ship, thought variable pitch propellers too vulnerable, said he would never climb to a crow's nest or let anyone else to do so in order to handle the ship, disliked anything but a standard pair of derricks to lift cargo or boats, and so on. Thus

---

37 Horizontal plating five times the depth of the frames, transversing them between decks.

the opportunity to lead the field by building in many of the developments up to that time, and providing us with an outstanding vessel was lost. Naturally money came into the equation and whilst Johnston's conservatism was welcomed by those responsible for the expenditure, the naval architects were disappointed at being restrained. Nevertheless she acquitted herself extremely well until the end of her life with FIDS in 1991, living up to her famous namesake.

Captain John Biscoe, 1794–1843

The ship and her predecessor of the same name were well known in the shipping community, and the origin of that name was well known in polar circles – but beyond them, sadly, hardly at all.

John Biscoe was born in 1794, and went to sea at 18 years of age when he voluntarily joined the navy. He saw service in the war against the United States, and at its conclusion transferred to the merchant service, serving as mate then master. In 1830 he was appointed master of the 148-ton brig *Tula*, which was to be accompanied by the 49-ton cutter *Lively* in an expedition in search of new sealing grounds for Enderby Bros, a well-respected ship-owning firm of the time known for its exploration for new lands as an adjunct to its commercial ventures. The abundant colonies of seals on the shores of South Georgia discovered by Cook, and those of the South Orkneys and South Shetlands discovered by Weddell and Palmer having been plundered to exhaustion by the many sealers that followed them, Biscoe headed for the South Sandwich Islands. He arrived there after revictualling at the Falklands, to find them barren of such spoils. Sailing south-eastwards as far as ice would allow in search of land, he raised it on 28 February 1831 in about 66 degrees south, 49 degrees east, and having sighted a bluff backed by a high mountain, named it Cape Ann, after his mother, then named the length of visible but unapproachable coast Enderby Land. The peak he saw beyond is now named Mount Biscoe.

Appalling weather amidst loose ice then nearly wrecked both vessels. The *Tula* lost one boat overboard and another was stove in, as were bulwarks and other parts of her upper works, yet whilst she survived the two vessels were separated and Biscoe feared that he had almost certainly lost his companion and the crew of the *Lively*, supposing them to have been overcome and foundered. During March he sighted more land, but with both his ship and crew in a pitiful state and continually fighting against more storm-force winds, he retreated and made for Hobart in Van Diemen's Land (now Tasmania). He arrived the following May to be met, purely coincidentally, by the great sealer Captain James Weddell. On passage, two of his crew had died, probably of scurvy, yet after some months of recuperation he sailed again for the Antarctic. Remarkably, in the mouth of the Derwent River on 3 September he met the *Lively* inward bound. Exactly six months had passed since their separation off Cape Ann, she having made Port Phillip.[38] There, she was taken from her crew by natives when her master, Captain Avery, and her small crew were seeking food ashore, and then reclaimed and sailed to Hobart by only her master, one seaman and a boy, all others having perished.

---

38 In Victoria, Australia.

On 10 October the two small vessels sailed yet again, this time north of New Zealand and thence south-eastward towards the South Shetland Islands, still in the hope of finding new lands and seal colonies. This was a further passage of remarkable undertaking, requiring a degree of seamanship, courage and determination that is barely conceivable, and which when completed meant that they had almost circumnavigated the Antarctic continent, their voyage mostly below 50 degrees south, and much of it below 60 degrees. Biscoe made his next landfall at the foot of the Antarctic Peninsula, then named Trinity Land, on the large island just offshore which he named Adelaide Island in February 1832 at about 67 degrees south, 69 west. He followed the coast north, he himself landing on the mainland, claiming it as Grahamland for the Crown before proceeding to the South Shetland Islands. Here the *Tula* was blown ashore and was temporarily abandoned but eventually refloated, and both vessels sailed for and reached the Falklands, from where they had sailed, completing their circumnavigation of the globe in the Southern Ocean.

The *Lively* was then wrecked amongst those islands whilst they again attempted to pursue their commercial interests, but in the end, with the crew of the *Lively* aboard the *Tula* but neither crew fit to do anything, Biscoe abandoned his searches and sailed for home, arriving in February 1833. He then settled to a more normal trading command, during which time he married, then he went with his family in 1837 to Sydney, Australia. After further ventures there, one involving sealing, he was overtaken by ill health and thus, unable to work, fell upon hard times. Donations were raised for his return to England, the subscribers being led by Sir John Franklin, the Arctic explorer who at the time happened to have become Lieutenant Governor of New South Wales. Biscoe sailed for England in February 1843, but never saw his homeland again, dying during the passage. His seamanship and husbandry of his vessel, it being so small and under sail, during those years south in tempestuous, ice-strewn and uncharted waters, and in freezing conditions must rank his voyage amongst the very finest, and him amongst the very best of British seamen.

Sailing under Captain Johnston

I spent five years as chief officer under Captain Johnston in the *John Biscoe* before taking command of her. They were seasons of much drama, during which my relationship with him varied from extremely good to poor. Together we experienced all degrees of excitement, concern, frustration and reward, he outwardly rarely showing much sign of any emotion, whilst I took each experience aboard, and gradually developed as a polar mariner. As often as I admired him for his skill I could have murdered him for his unreasonableness.

During the 1960/61 season, we attempted to establish a new base on the eastern side of Adelaide Island within the Laubeuf Fjord, but having spent a week hammering at unbroken fast ice that prevented our approach to the chosen site by some 20 miles, we turned our attention to the southern end of the island. The second choice of site was the only rock promontory protruding from the vast ice piedmont[39] rising inland from almost the entire

---

39 Ice covering a coastal strip of low-lying land backed by mountains.

shoreline. The new base was to facilitate the usual land and geological surveys of the area and to act as a meteorological station. Later it was also to serve as the operational centre for the Survey's newly acquired land-based aircraft, with an airstrip established on the nearby ice piedmont.

However, all lay separated from the deep-water approach by the Henkes, a three-mile barrier of islands, rocks and shoals through which there seemed little possibility of finding any route. Johnston repeatedly steamed the *Biscoe* east and west offshore in the clear water passage which is now named after him, eyeing the myriad of dangers that confronted us, together with the mass of loose ice amongst them. That ice ranged from large bergs showing that deep-water channels did exist, enabling them to get amongst the islands, to a multitude of smaller pieces, many aground and thus indicating some of the numerous shoals. He eventually chose to tackle the totally virgin ground with a direct course of 058 degrees towards the minute outcrop of rock where we hoped to land and find a site good enough to erect the huts of a new station. We successfully came to an anchorage in 20 fathoms, in a lagoon-like space between the main island and the offshore group, after feeling our way past several close dangers. This suited us fairly well, being about half a mile from the base site, except that we were continually harassed by bergs under way in each direction along our lagoon according to the changing tidal stream. In a later season over a two-month period we carried out a detailed survey of the whole area, more of which later, but it established a recommended approach course from the Johnston Passage with the base promontory on a bearing right ahead of 058.5 degrees – only half a degree different from Johnston's original run in.

Perverseness surveying

Once all the stores and materials had been landed, and whilst the Fids were ashore each day erecting the base, Johnston decided to utilise our time aboard by surveying part of the deep water beyond the Henkes. Even here we had encountered a number of isolated shoal patches rising abruptly from over 100 fathoms, which concerned us. We undertook several days' boat work during which we erected marks to fix our position on when running lines of soundings, and facilitate our approach to the base. We then took *Biscoe* to sea daily for about two weeks, sounding, fixing and plotting, but we failed to produce a worthwhile piece of work on account of the bergs that often obscured our limited number of marks to fix on, and caused too many deviations in what should have been parallel straight sounding lines to provide the correctly spaced coverage required. Early on in the ship's sounding we should have temporarily abandoned the plan and erected more marks with the boats, then resumed our survey, but Johnston would have none of it and pressed on. As the only qualified surveyor on board and his chief officer, it was me he took it out on for wasting our time on a useless survey. That resulted in a very strained relationship for a period.

Intractable conduct

As we were working many hundreds of miles of heavy pack ice with few leads towards Halley

Bay in the Weddell Sea, a further frustrating incident occurred. We had got into a pattern where Johnston handled the ship throughout the day and evening, occasionally being relieved by the watch officer, whereas I took over between midnight and 0800 hours. On this particular occasion I had begun to follow some loose ice and leads running south-westerly throughout the night, for there was water sky[40] ahead, indicating that this might be the best direction to follow. Although Halley lay some few hundred miles to the southeast, my advance southward was impressive. When Johnston came to the bridge in the morning he took one look at the ship's heading at that moment, saw my track chart of our progress and, despite more water sky ahead on the horizon, dismissed my night's work as useless. He turned the ship 180 degrees, cursing as he did so, and throughout that day retraced my hard-fought miles. At almost exactly where I had relieved him the previous midnight he got involved again in very heavy ice. After some further days negotiating it, we had worked our way through to Halley Bay. From there we could distinctly see water sky to the westward, beneath which we could almost certainly have reached, and utilised, the water found there.

The core of ice usually lay in the direction I had been heading in a few nights before, but at that particular time it clearly did not. Years later, with the benefit of satellite imagery, it was observed that a large polynia[41] would often develop in the unusual position of the centre of the Weddell Sea. Further, taking a south-westerly route towards what had previously been thought of as the worst solid core of the ice was often found to be beneficial. Fortunately for my disposition at the time of the reprimand, I had not the knowledge of later years.

Praise

On another occasion, the reverse. Late one season in the Gerlache Strait in fog with the occasional berg, navigating by radar at what we thought to be a moderate speed, with both Johnston and I on the bridge and a lookout posted right forward in the prow, we nearly came to grief. In the relatively calm sea the radar picked out both bergs and bergy bits for us to steer round, whilst small stuff was picked up visually and thus avoided in time. Suddenly a degree or so off right ahead there loomed a sizeable but completely smooth and rounded berg that had escaped detection by the radar. It had almost certainly been smoothed through its upper portion having once been under water then capsizing, presenting that surface to the air, its convex shape reflecting our radar waves away from our scanner rather than back to it. Johnston, on one bridge wing, having turned and stepped aft to shelter and light a cigarette as was his custom, failed to see it, and the lookout on the forecastle, fearing being in the way of the inevitable collision, ran from it without ringing the bell, let alone telephoning the bridge. From the opposite bridge wing, I saw the berg emerging through the fog, threw the telegraph to full astern and altered course to hit it head on, there being in my view no possibility of our missing it. Worst of all would have been to take a glancing blow on the shoulder or entrance where the hull fills out to its fullest breadth, and risk almost assuredly tearing an elongated

---

40 In a field of pack ice, a strip of dark sky on the horizon indicating the presence of open water.

41 An area of open water surrounded by sea ice, or an area of unfrozen water within the pack ice. It can be up to 200 miles in diameter, and can be formed by the upwelling of warm water which prevents the sea from freezing.

gash along the ship's side through more than one watertight compartment. It was better to stove in the bow and hope that the damage would be contained forward of the first watertight bulkhead. We hit it head on, crumpling the very eyes of the bow but no worse, although the shock throughout the entire ship was tremendous, with reverberations continuing for some time, as the bow rode up and down the berg in the swell, banging and grinding as it did so before any sternway took hold. There had been no time to tell Johnston and for him then to act, and I looked at him apprehensively as he then took her astern clear of the berg and sent me forward to inspect 'my' damage and to have the nearby tanks sounded.

Once the necessary review of our state had taken place he congratulated me on my action, the only time in five years I ever heard any praise from his lips, and embarrassingly, he repeated it both aboard and ashore for the remainder of the voyage. He was not always a difficult man, but nearly always keeping any emotions in check, he was both difficult to gauge and to serve. One never knew where one was with him behind that poker face, and I personally never gave of my best whilst being overseen by him.

Resultant relationships

Relationships aboard apart from those between the master and his deck officers, who found him cold and distant, and between him and myself, which waxed and waned as described, were generally agreeable. Johnston had only one buddy aboard, the chief engineer, Herbert Ward, an ex-Royal Navy chief engine room artificer, of about his own age. When passage-making they would often play cribbage together, seated beside the wardroom fire with a gin to hand. 'Digger' Ward was a grumpy, unhealthy-looking man with a heavily lined face who chain-smoked. He told the most outrageous stories of his involvement during the war, that put him in almost every theatre of action simultaneously, and the younger members of the wardroom teased him mercilessly. He was a man of habit; each day after a spell in the engine room, he took an afternoon nap, followed by time in the steering flat or checking deck machinery, then his pink gin before dinner, and toasted cheese late, before turning in. His tormentors filled the gin bottle in the wardroom liquor cabinet with water immediately prior to his six o'clock tot, which he never noticed – and they put a penguin in his cabin during his siesta and soap in the pantry fridge from which he made his welsh rarebit, both of which he very much did!

He was, however, an excellent engineer. On every emergency occasion when he was required to start up the main engines, he managed to do it within a minute of being called, wherever he was. I do not remember one main engine breakdown, but I do remember the praise and respect he received from consultant engineers. He was fiercely loyal, with duty as his priority; I was extremely lucky to inherit him when I took command, and he threw his support behind me. However during those early days of my being chief officer he regarded me as too young, and conveyed some doubt as to my ability. Understandably, for my predecessor David Turnbull, was 13 years my senior and had already held command, although I do not remember concerning myself with that at the time.

Amongst other officers and crew, work, whether it was the routine running of the vessel, preparation for the south or tackling the extreme circumstances of those southern voyages once there, dominated most of the shipboard relationships. Normally there was little time for discord, and there was always a heightened camaraderie in general adversity. It was a very busy, professionally run, serious ship, without too much humour, but neither was there disharmony. Down south, the pace of progress round the bases, the work at them and the surrounding interest of everything polar, be it scenery or wildlife, and for the deck department the passage-making and surveying, left little time for anything but accord. Paramount, though, in maintaining the attention of all aboard a great deal of the time south, and bonding everyone together in harmony, whether in fascination at the conditions or self-interest in survival, were the elements and the regular fight against them.

Ordeal by wind

During one of my first seasons in *Biscoe* we made a late call at Hope Bay at the northern tip of the Antarctic Peninsula, an anchorage with a reputation for a bad combination of poor holding and strong winds. As was our custom after a day's work, the majority of base Fids came aboard for their evening relaxation. At around 2230, as their time to leave approached, I went on deck to have the workboat swung out; it had been lifted clear of the water to remove it from the danger of ice passing down the ship's side. Many of us were in shirtsleeves, for despite a light wind it was remarkably warm, but by the time the Fids had boarded the boat a stiff breeze had sprung up, and when the boat returned from the shore, we hoisted it back aboard in about 40 knots of wind. The anchor cable began to jar as our yaw increased, then instead of doing so rhythmically either side of a central position, the vessel began to hang at the full extent of each yaw almost broadside to the wind.

We well knew that the problem about to arise on such poor holding ground would be the anchor dragging. So starting the main engines we laid out a second anchor up into the wind and rode to eight shackles of cable on each, they bearing equal strain in the form of a wide V shape. Fairly quickly the wind strengthened still further and with both cables gradually drawing straight ahead it was clear that we had started to drag, so decided to pick up altogether and leave the bay for the safety of open water. The windlass found it hard going as we tried first to lift both cables, but trying them singly made for slow progress. Steaming ahead into the eye of the wind to ease the strain on the cables, we invariably fell off the wind as the cable became vertical or as we attempted to carry it forward to hold our line. The needle of the anemometer was now reportedly banging against its upper stop as the gusts exceeded 120 knots, and we on the forecastle found it hard to breathe, see or hear. Each time we fell off the wind, unable to maintain our heading, we dragged further shoreward. No matter what Johnston did, even for a while reverting to lying to nine shackles to one anchor with the other on a shorter scope underfoot in conjunction with engine movements to prevent the excessive yawing, the combination of those irregular squalls of hurricane-force winds, the poor holding and our own light ship profile lost us the battle, and just before midnight in pitch dark we were driven ashore. First the starboard quarter hit the rocky shoreline and then, the anchors still slipping we fetched up broadside on, holing a ballast tank and an empty fuel tank beside the engine room.

The circumstances that brought about these sudden horrendous conditions are interesting, as is the manner in which we dealt with our ultimate predicament. Hope Bay is located on a deeply embayed, fragmented coast, backed by high and steep mountains. When the wind in a depression aligns with the valley at the head of the bay between two adjacent peaks, a sudden onslaught of strong, gusting winds can be created in the form of Föhn winds. The term originated in Alpine regions to describe these phenomena, also found in North America as the Chinook, and in Argentina as the Zonda – but few of these, I suspect, have the polar ferocity of a Hope Bay Howler.

With the passage of a depression across a mountain range the moisture-laden air is forced up the windward side, releasing precipitation as the atmospheric pressure drops, and the air both cools and dries. Orographic cloud often builds close to the summit ridge, then pours over it, dispersing as it descends on the leeward side. Lenticular clouds can also be formed above the land, detached from its highest levels. As the then dry air descends it warms again, but because the rate of warming of dry air is greater than the rate of cooling of the original moist air, the air to the leeward of the mountains is much warmer than that to the windward at the same altitude. It gathers speed down the leeward slopes, and the resultant wind often accelerates yet more, as in the case of Hope Bay, by being funnelled through gaps in the mountains. It also becomes disturbed by the irregularities of the terrain over which it passes to produce its gusting nature, that too having been so on this occasion and contributed to our downfall as successive blasts caught us broadside on.

After the concern of actually grounding, something we had all felt capable of avoiding despite the ferocious conditions confronting us, and the unease that came with having failed in the seaman's main objective, which is to remain afloat, we took stock. We confirmed the extent of the damage: neither any double-bottom tanks nor the main hold were ruptured, and the chief engineer reported that the engine room was intact. We remained almost upright, and within the hour the wind eased, but not below 60 knots and still gusting to 80 from exactly the same direction as when we had fought it to stay afloat. Several times Johnston went out onto the leeward wing of the bridge with his cigarette, as always, cupped in the palm of his hand and peered overboard into the darkness of rocks, brash ice and swirling water but could make out little of any consequence. Finally turning and coming within the wheelhouse, he announced in his Northern Irish brogue 'We'll have a look after breakfast,' and after an aside to me, 'Keep an eye on her, Tom,' he descended to his cabin. It was yet another example of his imperturbable nature. No regrets or comments as to how we had got into that situation, let alone painful analysis, just calm acceptance of the present situation and sensible recognition that in the dark with an onshore storm-force wind, yet without swell and insufficient sea running to produce any movement of the vessel, let alone pounding or grinding against the rocks which might damage her further, we could not and should not try to do anything. Incredibly, though, we had been told that we would appraise the situation not at first light but after breakfast.

It being spring tides it was deemed imperative to attempt to refloat before we lost the advantage of the good highs. Therefore despite onshore winds of 40 knots the next morning, having ascertained that the propeller and rudder were free and apparently undamaged, we attempted to haul off on the rising tide, only to find that as soon as the bow angled into the bay we blew back again onto the shore, dragging home our anchor and an ever-lessening length of cable. We still had the shorter length of cable and second anchor dropped underfoot, though then leading slightly into the bay, for this too had been paid out somewhat further just prior to the final stages of grounding. We decided to lift our spare anchor from the well deck and carry it out as far as possible on a mooring rope that we would attach to that shorter length of cable.

Working on the weather side of the ship, the leeside being inaccessible against the coast, we prepared the scow with the anchor atop some deals and with the mooring rope, shackled to it, stowed beneath. Then we attached the other end of the mooring rope by way of a strop around the shorter cable, in the expectation that it would work its way sufficiently down towards the anchor. With motorboat and lifeboat towing the heavily laden scow, and in fear of sinking the lot into a 'buggers muddle' we set off upwind. I was minded of classroom discussions whilst studying, with lecturers expounding methods of seamanship, and I wished that some of the inexperienced academics could have been in those boats attempting to work upwind in those conditions for several hours. In fairness, though, most of their teachings had sent us to sea with sound basic training.

With little freeboard, solid freezing water lopping on board threatened to sink us, as the mooring rope dragging astern curtailed our ability to keep our boat's head to wind. Freezing spray lashed our faces and encrusted the mooring rope before we had time to pay it out, and both spray and rope numbed our hands as the gunwales, side decks – indeed everything bar the engine casings – became covered in ice. The engine casings provided the one haven in each boat where we could occasionally warm our hands. Throughout our progress, I had my eye on those wooden deals, with my hand poised over my knife; were our entire floating circus to succumb to the elements, then the deals, once cut away, would float and would hopefully, together with an upturned scow and partly submerged lifeboat, support us and enable us all to reach the ship immediately downwind before we fell victim to the freezing water. It must be emphasised that two thirds of those aboard that detached flotilla were Fids; scientists, general assistants, diesel mechanics, cooks and the like, not seamen, none of whom had signed up to be near drowned, yet they did as they were bid incredibly well, and I felt very responsible for them. Having paid out as much rope as possible, using the workboat's echo sounder to locate a deep gulley as we went, we managed to tip the anchor overboard in a place which would improve its chances of holding when hauled upon. No sooner had we rid ourselves of the anchor than Johnston recommenced heaving. As we returned to the ship, she was already separating from the shore and canting seawards, providing us with a lee from which to board. We retrieved our tackle, re-anchored and recommenced work. The camaraderie which ensued from this near-disaster turned into triumph, involving so many of us, both Fids and crew, was euphoric and lasted from then until the end of that season.

In later years when at anchor working that base, and the temperature rose significantly in fairly calm and pleasant conditions, I would stop the resupply, have all craft hoisted and stowed, and much to the amazement of most aboard, start the engines and shorten cable; as the wind increased in the bay I would take the ship a couple of miles offshore. As we distanced ourselves by just that small amount from the coast, the wind would noticeably decrease, which showed how local was the phenomenon. Many aboard then marvelled at my apparently psychic qualities, but for me it was by then pretty simple stuff: 'Warm wind – get out!'

Ordeal by ice

During a season of severe ice in 1963–64 we had worked the *John Biscoe* through increasingly heavy and tight pack ice as we made our way southwards outside, that is to the west of, the Biscoe Islands. These lay to the west of the peninsula, between 65 and 66.5 degrees south, our destination being the Adelaide Island base. The pack had almost certainly originated in the Bellingshausen Sea to our south-westwards, and coming up against those offshore islands it was halted and began to compress, changing from an easily navigable, wind-driven loose pack of five-tenths sea coverage, to one of almost ten-tenths, requiring a great deal of forcing and breaking. In these circumstances it is essential to know when not to overcommit the vessel and to judge when it is still possible to turn and retreat to the easier conditions, providing the freedom to choose when, where and how to make another attempt at one's destination. On this occasion this was completely misjudged, although some exoneration must be made, for the wind that was driving the pack shoreward never let up.

The ingrained philosophy on the bridge was never to fight the elements too hard or unnecessarily. Absorb or ride their punches, or when necessary retreat if possible, and be mindful that most harsh conditions change, and with time and patience they will eventually alter favourably. Yet this time for several weeks the wind blew constantly onshore, from between southwest and northwest, sometimes imperceptibly, sometimes strongly but always westerly. Furthermore, we had not been aware originally that a vast field of pack ice lay out to sea to our north and west, although latterly in the period of becoming beset there were hints of this further ice: there was a lack of swell entering the ice around the ship from the supposedly open sea to the north-west, which could only mean that it had been dampened by ice; ice was also indicated by ice blink, the lightening of the sky from the reflection of light from pack ice, or indeed snow-laden land or an ice shelf, up onto the clouds above, seen at distances far beyond that of the horizon. It is very distinct from water sky. The significance of this further ice to the north and west was that driven by the wind, it curled around to close in behind us, also becoming compressed against the islands and making any escape northwards impossible.

The severe conditions there meant that working south was just as easy if not easier, so risking the unlikely possibility of becoming beset in January right through and into the ensuing total winter freeze, usually commencing at the beginning of April, we attempted to forge south towards our destination. For a week or two we started up the main engines each day, often leaving it until about four in the afternoon if the sun had been shining, to provide some warming of our hull during the day and hopefully slightly melting the immediately adjacent skim of ice. The resultant progress would sometimes be measured in miles but more often in yards, or even, with incredible patience but the use of precious fuel, in inches. To those without knowledge of working pack ice that may appear to be futile, but despite the pressure and apparent uniformity of surrounding sea ice, within vast areas of it there are often differing masses depending upon its origin, formation, and how it has travelled and mixed, to where it is met. For instance, much heavier floes will hang back to windward in loose conditions because of their draught, whereas the lighter ones, sitting less deep in the water, will race away downwind. Then those larger floes will, when they collide with each other, sit awkwardly, one protrusion against another, leaving areas of water between them. Heavier floes are therefore often easier to work than lighter ones all squashed together. Utilising that water, and chipping away at their points of contact, can be a successful method of progress in such conditions. Therefore to inch one's way forward in the hope of coming upon more workable conditions makes sense. We also had an as yet unspoken agenda, for we were carrying nearly 80 huskies, transporting them from the northern bases for their introduction to the Adelaide Island base. We had been getting sun sights and were able to monitor our progress; it had become apparent that we were moving south within the ice much faster than warranted by our own efforts to get through it, and were drifting south across the entrance to the Matha Strait.

With such advancement, the northern tip of Adelaide Island was soon to be on our radar. Remembering that our purpose was to relieve the base there, rather overshadowed at times by our own wellbeing, Johnston and I held discussions with the experienced Fids aboard, and together talked of a modified relief. A plan evolved to deliver the essential stores and exchange personnel by dog-sled, firstly across the sea ice to the island's coast, thence down its length just inland on the piedmont to the southern tip where the base was situated. As the island's coastline of continuous ice cliff, rising piedmont and high inland mountains hove into view, we became beset. Not surprisingly, for any filtering and escape for the wind-driven pack between the Biscoe Islands or through the Matha Strait was now denied, and the pressure increased as it came against this long, unbroken coast. By taking bearings of the land we soon noticed that we would drift north and south within the ice on the tide, but that our southerly advance was at least quadruple that of our northerly. There was, as well as the tidal flow, a very strong southerly current that began to funnel us between the main island and an off-lying area of islets and shoals, the Amiot Islands. This area was also littered with innumerable grounded bergs. The additional compression that this funnelling effect caused pressure such that at times we were canted over some 15 degrees. Ships of earlier times built to withstand ice had barrel-shaped hulls, which when under pressure in ice assisted the ice-flows to slide beneath a vessel and raise her, ultimately in theory for her to sit on top of the pack. But ships were smaller and lighter then, and although many modern icebreakers have an expensive-to-build curvature of hull to reduce pressure when beset, we were slab-sided, and the force and build-up of ice on the weather side heeled us over. Despite static conditions such that we remained precisely within the same floe complex, surrounded by our footprints, those of the dogs being exercised, and the inevitable fouling and trash – minimal, for we disliked sitting amidst our own refuse – we continually changed our heading, to such an extent that both port and starboard sides took turns, during the weeks that followed, between facing westward and landward, with the ship heeling alternately to port and starboard.

During such a prolonged period of reduced activity it benefited most on board whose normal tasks were unnecessary for them to be found work. So the deck crew repainted the accommodation until our store of paint ran out; engineers overhauled auxiliary machinery, for we had eventually given up the daily attempt at moving, and surprisingly the main forward 'tween deck also became a hive of activity. We were carrying two partly dismantled de Havilland aircraft, with the engine and airframe fitters to maintain them also aboard, the latter being keen to be put ashore and get on with their real job. By way of a substitute, they began work on the aircraft and engines, but to do so they had to be kept warm in the unheated hold. Amongst their equipment for use ashore they had a giant Nelson Norman fan heater that raised the temperature sufficiently to enable intricate work to be done with bare fingers. Subsequent discussions regarding what followed revealed that slightly unusual noises had been heard by those in the 'tween deck, though not significantly different from those that we heard, and were used to, when in ice, to cause comment. Down below, in the hold in particular, when working ice there could be a great deal of noise as it banged and scraped against the hull; in fact it was educational to send young deck officers down from the bridge into the hold, for it could be quite alarming, and thus helped temper their subsequent handling of the vessel. When beset, without the main engines running, there were creaks and groans as the pressurised ice moved against the hull, even sharp retorts as the ice outside shattered. On this occasion, sporadically amongst those eerie sounds there were similar noises, their significance un-noticed, as the ship's side frames cracked and split where upper and lower sections were joined to form one continuous member. The theory expounded later was that it was the temperature gradient across the steelwork of the hull, from well below freezing externally to plus 15 degrees centigrade within, in addition to the strains brought about by the pressure, that had caused the welds to rupture by the dozen. It was an interesting although slightly alarming sight to see the ship's side plating only slightly distorted and indented, yet many of the supporting frames split horizontally and separated by some millimetres. The thought of the whole ship's side caving in was hard to ignore.

At the time we discovered this damage we heard on the radio that the Argentine icebreaker, the *General San Martin*, was not too distant, beset near the Amiot Islands, and also in trouble. Usually maintaining an unfriendly silence, they had agitatedly disclosed their plight of having ruptured a fuel tank that contained aviation spirit. This had leaked, and was frighteningly evident for it had stained pink the floes trapping her; her crew were going about her steel decks in socks, for fear of creating a spark. Thankfully we always carried our avgas on deck in 40-gallon drums.

The enforced inactivity on board had mostly been compensated for, and morale remained good although there was an amusing line of thought quietly running beneath the surface within the crew. As always the psychology of our life was interesting. During the first week or two of being beset, a young mess boy and then a junior rating came to see me with a reason, such as a sick mother, for him to be the first to be lifted off were any assistance to come our way and evacuation to take place. This then progressed upwards through the ranks of the ship's company, arriving at my cabin door with more detailed requests, until it was the turn of relatively senior officers; they explained that having knowledge of our predicament and exactly where we were, they should be taken off first in order to help organise any subsequent relief, it being, they said, the duty of the captain and mate to stay with the ship in the case of abandonment! There was talk of American icebreakers and helicopters appearing out of the blue, with complete disregard to what was known to all – the poor performance of such vessels in these conditions, and that nearby was an icebreaker beset under severe pressure in a worse state than us.

As the weeks passed our southerly drift diminished, and the dogs were spanned out on the ice, the hold opened, sledges prepared, and cargo sorted in some order of priority for delivery to the base. Our emergency plan for the relief was put into operation. On a near-calm day in sunshine it was an exhilarating sight and sound to watch the huskies commence hauling their sledge loads to the coast, for they were ecstatic to be off the ship and working again. But it was slow going, with a five-mile traverse of the chaotic sea ice under enormous pressure, with much rafting, heaped ridges and tumbles of fragmented blocks to be overcome. Then they had to find a route up the ice cliff onto the crevassed piedmont before a journey, in excess of 25 miles, to deliver their loads to the base. It was hard work ashore but very easy aboard, discharging just a couple of sling loads of cargo per sled, then reloading and securing them as the teams arrived back alongside. Whilst ensuring that everything at our end of the exercise ran smoothly, we also fed the dogs with seal meat from the carcasses we had had on deck, having taken them as opportunities arose during our time within the pack.

One morning at 0400, this leisurely shipboard existence exploded into a flurry of activity. After several weeks of an onshore wind pinning us within the pack against the coast, it changed direction to blow strongly from the east, offshore. We scrambled to get the dogs and equipment on board. By 0600 hours the ice was easing around us, though not those floes immediately alongside. We remained firmly in the grasp of the same floes that had lain beside us for weeks. Soon it was apparent that within those floes we were well under way; the pack was recoiling seaward like a giant spring, taking us with it. With engines running, propeller turning and rudder free we remained beset, but were travelling in excess of 12 knots; the simple method of a radar distance to a static berg and a stopwatch gave us accurate speeds. The engines were put alternately to full ahead then full astern for some minutes each, as the rudder was ranged from side to side in an attempt to free the vessel.

We realised quickly that static bergs, whether aground on a shoal or stationary in the water because of their great depth, presented not just a navigational aid but also a great danger. Many times when working pack ice we had witnessed floes carried by wind and current streaming against such bergs, piling up against them and being smashed to smithereens. Then the dense debris would heave skywards under the enormous pressure and slide around the mighty obstacle and a vortex of open water created constantly downstream of the berg, until the pack closed together again. We had passed many such bergs smashing the pack within a hundred yards of us, only this time we were not working the pack, but were stuck in it and being carried at its will. As we prepared to lift the motorboat from the foredeck to swing it from side to side in the hope of rolling the vessel, and to rig fire hoses to pour water down the ship's side in an attempt to break the friction between the hull and our imprisoning floe, we realised from our observations that we were travelling in excess of 14 knots broadside on towards a cluster of grounded bergs, and that one of these lay directly in our path. Unable to alter our heading, and remaining beam on to it, we attempted to inch ahead or astern in the hope that we would draw clear enough not to slam into it amidships. It made no difference. Reminiscent of photographs of the men aboard Scott's *Discovery* poling ice away from the ship, many appeared on deck, without instructions, with oars and pieces of dunnage[42] to fend off our ship from imminent disaster in true British fashion. The pack, as we bore helplessly towards the berg, was a sight to behold, first smashing into it, then being shattered and ground into minute fragments. Was that to be our fate? Still broadside on to our line of advance and at great speed, amidst the thunderous sound of disintegrating pack ice, we miraculously skewed away from it, our stern clearing the berg by about two feet. As we passed it we heeled violently over from the increased pressure, before becoming free of our then-fragmented captors. Once we were in the wake of almost open water that forms behind a berg in these circumstances, as if it were itself steaming through the pack, we took control of the vessel for the first time in weeks. Then re-entering the pack we began to make good the course required, south and around the southern corner of Adelaide Island towards the base, which we then reached within 12 hours.

Adelaide Island, the Argentine Islands, Deception Island and Halley Bay became the main locations of our work during those years of my being chief officer, and we also visited Stonington Island in Marguerite Bay and the northern bases of Port Lockroy and Signy Island.

Adelaide Island

The Adelaide Island base was established during the 1960/61 season after two years of bad ice had made it impossible for the ships to sail beyond the Antarctic Circle and therefore to visit the bases at the foot of the peninsula and an intended site there. Late in that season, much easier ice enabled us to enter that southern region, but still left much fast ice in areas protected from the weather, particularly the swell, with the result that our intention of establishing a new base on or near the north-east coast of Adelaide Island had to be abandoned. After more than a week attempting to break a canal, having made only a mile into a 30-mile expanse of fast ice towards our objective and with no sign of any assistance from nature, we capitulated and turned our attentions elsewhere. Also we were now mindful as to whether the proposed site was a sensible choice were such ice conditions to prevail in subsequent years. The promontory we retreated to was at the southern end of the island; I have already described Johnston's first exploratory approach to it. The base had initially been intended to support the standard survey and geology of the island, and to establish and support a small advance base at Fossil Bluff in King George VI Sound, 250 miles south across Marguerite Bay, but during the first two years' tenure there it became apparent that the ice piedmont rising behind and above the rocky outcrop could serve as an airstrip from which survey parties could be flown to work

---

42 Rough timber used to separate and assist the stowage of cargo.

much further afield. This location and the surrounding area, being of great interest, had been discovered by Fuchs when sledging from Stonington Island in 1948.

The two aircraft, a Beaver and Twin Otter from de Havilland, had been re-introduced in 1959/60, no flying having been undertaken since one season in the late 1940s. Staging in from Deception Island, where they were based, they sometimes landed at the Argentine Islands en route. Regrettably on one such visit in their second season the Beaver, piloted by Ron Lord, an outstanding but extrovert RAF flight lieutenant on secondment, landed on the frozen creek beside the base but then, moving aside to allow his compatriot, Paddy English, another RAF flight lieutenant flying the Otter, to land, the Beaver went through and despite desperate efforts by all to save it, the plane was lost.

The succession of pilots from both civilian and service backgrounds, some flying in from South America having staged all the way from Canada, others commencing their flying off the water or ice down south, deserve much fuller mention than I shall give them here, for my story is primarily one of ship work. I am likely only to mention those 'fly boys' in passing, and then mostly only when we were involved with them, that usually being during incidents (a euphemism for accidents). They were extraordinarily competent in very difficult conditions, courageous, and above all never an elite separate unit but so much part of the overall team, never denying anyone their co-operation or support, and many times flying me to observe coastlines and ice.

On a voyage to this base their well-intentioned assistance did not result too beneficially for us; in fact it created a potentially critical situation that fortunately ended only in humorous delay. Meandering through ice to find the easiest route often took us near the coast where shallow water and grounded bergs broke up the edges of an ice-field, and offshore breezes caused by cold air falling down the slopes of large land masses eased the pack seaward. On this occasion, taking a passage close inshore to reach the base, we were assisted by the aircraft that had already arrived there for their flying programme; they became airborne to help us find a route. We had worked to within half a mile of the shore where water was found, when the two aircraft appeared at low level and made a pass, and then returning at barely above mast-top height they bombed us with paper bags of flour. When the amusement was over they guided us through a great many bergs to places that they could see were free of shoals. But a problem arose in that they led us from one patch of open water to another between two icebergs, only for us to find that they were one; the two well-eroded high pinnacles stood apart, with just room between them for us to pass through. As we entered the gap, too late to alter away or stop, we saw that the bergs were joined beneath the surface. Fortunately that submerged portion was well rounded from much weathering, and we rode gracefully up onto it with little impact; there we stayed for several hours as if in dry dock. No amount of engine or rudder movement would dislodge us, yet while we were having dinner tilted to a very noticeable degree, bow well up and stern lying deeper in the water than was safe, the ship suddenly slipped astern to regain her equilibrium and we recommenced our passage. The lighter side of the episode was enjoyed, but in truth we had been concerned throughout that our bearing down on the underwater portion of the berg might cause it to crack and in separating it would become unstable, those towering pinnacles then to fall on us.

Immediately after our initial successful approach passage we were confronted with the formidable task of landing anything at all. It was an uncomfortable anchorage, with bergs trapped between surrounding islets and reefs harassing us and the boats alongside on each ebb and flow of the tide; although many of the bergs were higher than the bridge and passed us four times a day we were used to dealing with the problem by handling the cable and veering away from them. Worse, though, and exacerbated by any swell, was the conglomeration of very heavy hard glacial ice from the nearby ice cliffs that collected at the landing area and rarely left it. The motorboat and scow had to be eased through this barrier, often 100 metres thick, as it seethed dangerously around them. Handling the boats in the ice and achieving a landing on ice-covered rocks amidst deep gullies full of water, was bad enough in itself, but the craft were pummelled by ice once they lay alongside the rock face, and the buffeting they received made the task of offloading, particularly large items such as generators and tractors, particularly hazardous. Ashore, slippery slush from continual conveyance of stores over a narrow route made for further precarious work.

In the later season of 1962/63, an attempt was made to fill some of the gullies with rocks, cement them over, and square off a face that the boats could lie against, although little could be done about the continual presence of the floating ice. To this end one of the builders engaged to enlarge the complex ashore had knowledge of blasting, and we carried explosives for this purpose. Already nicknamed Bang by the crew, he set about his task, and with offloading halted and onlookers warned away to a safe distance he set off his first explosion. As the smoke cleared Bang was seen to be lying knocked out, his head bleeding, having been hit by a piece of his own handiwork. He recovered to finish his work, though the onslaught of sea and ice reduced the construction to nearly its former natural self within a season.

At this landing after the season of weeks' besetment, the problems were compounded when Johnston became concerned that we might now become beset within the anchorage. Much pack ice lay to seaward, and with dropping temperatures the season was drawing to a close. The landing had on this occasion become totally debarred from use by ice, much of it the size of houses, bound solid by brash ice. A very poor alternative was found; rocks not much above sea level, backed by an ice cliff which all stores had to be lifted up. The stores having been piled ashore so urgently, the ship departed before the base personnel had time to move them up onto safer ground. Bad weather came in and snow drifting seaward off the cliff covered and then froze them to an irretrievable depth within days of our departure; 80 40-gallon drums of avgas, an aircraft propeller and a spare set of skis plus about a ton of anthracite fuel were lost. The following season, flying hours being in jeopardy because of that loss of aircraft fuel, I (by then in command), was asked to consider visiting the base early in the hope of replacing the fuel to get the flying programme back on track. We arrived, with little difficulty from ice, two months earlier than in the previous season; such were the vagaries of sea ice cover from year to year and the flexibility required of our itineraries to cope.

Despite the pack ice often severely obstructing our passage to this station, the poor landing and the difficult haul up the slopes of the piedmont to the air strip, where the combination of summer temperatures on the snow cover and the repeated heavy use of the track bogged down muskegs and sledges en route to the landing strip, the Survey persevered with Adelaide as a base. After two years the topographical survey and geology of the island had been completed – and although the landing strip was not ideal, white-out being a problem which caused one aircraft to crash-land and, abandoned beside the runway, become a permanent reminder of that danger – the successful extensive support of field parties and likewise the sustainment of Fossil Bluff way to the south, warranted the continued existence of this base.

A major survey

On account of this it was decided that we should dedicate January and February 1963 to surveying the anchorage of this base and its approaches to RN Hydrographic Survey standards. We had embarked a naval survey party under the command of a hydrographer, Lieutenant Commander Barry Dixon, with their launch and a copious amount of equipment, for they were also to camp and work with me from ashore. His team included a petty officer and several ratings from the Hydrographic Service, and four Royal Marines to support and assist, all of whom very quickly integrated with our crew. Their sturdy survey launch with Kitchen bucket steering gear, ideal amongst ice, and with an ample closed area for plotting, was stowed on deck. Barry had spent a season in HMS *Owen* surveying the western end of South Georgia, and was therefore fairly familiar with the severity of the weather likely to be encountered, and the need, whilst being enthusiastic and ambitious, to temper those characteristics with caution in these waters; he soon became an amiable member of our ship's company.

The first task was to establish survey marks, and between two of them, relatively close together, a base line was measured, then the whole area triangulated. This entailed a great deal of boat work, utilising the naval survey launch, the *Biscoe's* workboat and our inflatable craft, ranging from the ship each day. The ship was repositioned each evening to facilitate this ground work by reducing the steaming the boats had to undertake to erect the marks, but most sites were still some distance away from the ship and had to be visited more than once to obtain the required angles. This resulted in long hard days under way getting wet and cold, and the same, with only slightly less effect from the weather when ashore, patiently setting up and using the theodolites. Late in the day we would return to the ship for dinner and resolve the calculations of the triangulation in the warmth of the chart room whilst the ship moved to the next location to find an anchorage.

Barry, like most naval hydrographers, was enormously keen and diligent, and had a knack of keeping everyone with him working hard. For me he became a very pleasant companion – but the two boats' crews took to whitewashing his cabin window to lead him to believe that there was fog in order to prevent him from starting too early in the morning! As the survey progressed, as expected the ship's role of sounding offshore progressed much faster than that of the boats, both sounding, and fixing[43] rocks, shoals and islets. The *Biscoe* therefore departed to support activities at other bases whilst Barry and I set up camp ashore on Avian Island, only

---

43 Establishing the position of the ship, boat or a feature.

two cables off the main island. To improve the output of this shore party, I took the *Biscoe*'s workboat with me, with some deck hands and an engineer to man and maintain her.

Avian Island, about half a mile square, rocky but ice-free, rising to no more than 140 feet, had at that time only a few patches of snow lying in south-facing hollows. It provided an ideal site for a camp to establish ourselves in. The island also had a well-indented coastline, which afforded a choice of long, deep, and narrow creeks in which we were able to moor our two boats centrally with lines ashore from all quarters, yet have easy access to them. We spanned the entrance of our chosen creek with a spare length of anchor chain to prevent bergy bits and brash from entering and damaging the craft.

The most significant aspects of the island for us were its suitability to house the camp and boats, but as its name portrays, for others it was the summer nesting bird population that was its most notable feature and attraction. Siting the camp had not been easy, for Adélie penguins, the most belligerent of their species, occupied almost all the flat ground, and giant petrels, cormorants and skuas the majority of the remainder; these birds were delightful to visit and observe, but difficult to live amongst. Our marines were splendid in every aspect, likeable, cheerful and hard-working, and they ran the camp superbly, whilst nothing extra was ever too much trouble for them – not even dealing with the skuas that continually attacked our heads, particularly if we wore a hat made of anything resembling fur. About the camp and particularly when we were going to or from the boats, invariably carrying something and unable to defend ourselves, they would dive-bomb us. We observed Sunday as a rest day for the team to relax whilst Barry and I spent some part of it working up or plotting our findings. Possibly not to be applauded nowadays, the Royals chose the dispatch of some of the more aggressive skuas for their Sunday morning R and R. A mighty fusillade of shots from out of sight broke the silence, to be followed by the reappearance of our guardians with one of their number carrying just one dead skua. As these remarkable men were known for their compassion as well as their toughness, resolution and professionalism, we wondered if they had actually balked at their self-set task, instead merely engaging in some inanimate target practice. They were, nevertheless, ribbed about it endlessly.

Their devotion to duty at all costs was exemplified by the actions of two of their number whilst rock-hopping. It was our practice to fix small rocks and islets that were almost awash by taking the motorboat alongside them and both myself and a recorder jumping out, I to take the requisite sextant angles and my colleague to note them in the record book – this was rock-hopping. An appreciable swell could make this job difficult, and on one occasion so much so that we were both washed off; Sergeant Worth, in charge of the marine contingent and acting as my recorder on that day, could only be seen by his extended arm and hand, keeping the record book dry. The surge around the rocks made it difficult to stay above the freezing water, and indeed dangerous, but the brash ice swirling against us made it perilous. Fortunately after not too long we were brought aboard, but only after a second Royal had jumped in to assist us. That group of rocks now bears the name Worth Reef. All the names of those who contributed to the survey, as proposed by Barry, were accepted and appear on the chart today. All historical matters and associated names have to be taken into account when naming features, but here there were extremely few, indicating how virgin these waters were. Captain Johnston had the main western approach passage named after him, I the eastern approach channel and the route into Marguerite Bay, the chief engineer the significant Ward Rocks, the third engineer who maintained the boat engines Glover Rocks, and so on right down through the ranks, a delightful reminder of all those who took part, too many to detail each individually except for one final episode concerning the naming of an island. Tommy Biggs was the hard-working, reliable but somewhat deaf Falkland Islander coxswain who I took ashore with me from the *Biscoe* in our workboat. In wind with a high degree of engine noise, precise clipped orders were given to the coxswain from the cabin where plotting was conducted to maintain the chosen course, and similar from the open cockpit when conducting sounding runs, rock-hopping, or fixing rocks when afloat as close as possible to them. Hence this style of conning was used as we approached a small islet in far too large a swell to be attempting either a landing or staying close to take angles. With little time to spare, I saw my folly in continuing, and shouted, 'Full ahead and hard a'starboard!' to clear the danger. But Tommy heard only the 'Full ahead'! We rode onto the islet that was then more like a breaking rock. The sea fell away and we were planted high and dry, encircled by seething white foam sucking away beneath us, until a later swell lifted and carried us over the top and beyond. Hence one of the Falkland Islands' best known surnames appears on that chart as Biggs Islet.

During our time encamped ashore we were visited by two ships during what was materialising into a very ice-free season. The first was HMS *Protector*, the Falkland Islands guard ship with the more recently acquired, somewhat overblown title of the Navy's Ice Patrol Ship, making a far deeper foray south than usual, whilst the second was an American Wind Class Icebreaker. I had come to know *Protector* and her officers well. From all accounts she was far from the best sea ship, and her large hangar aft, added to accommodate her helicopter, made her very unwieldy; worse, despite reinforcements to her bow and other modifications to her hull she was not very strong. Visiting Adelaide Island and intending to enter Marguerite Bay, both ship and captain were far beyond their sensible limits of responsible operation.

She was Barry's parent ship, so I joined him on the morning of her arrival to go aboard with our party for a wash and brush up and to collect some additional survey gear, and for Barry to brief her commanding officer, a four-ring captain, of our activities and progress. Barry brought her into the position of the best anchorage, but her CO insisted on dropping the anchor in the deep water between the main and off-lying islands, farthest from the surrounding dangers. Possibly sensible normally, but in this case not a position we would have advised, because of the risk of dragging in the frequent strong winds in the deeper water and the likelihood of being harassed by a higher number and larger size of bergs continually under way there. Our party were all in a state of undress when the klaxon sounded and announcements were made over the tannoy system to close all X, Y and Z doors, the ship was in danger of being hit by a berg, and they were weighing anchor to put to sea. Barry, fearing that after that early alarm they were unlikely to return to the anchorage to land us and that being flown back to our camp in their helicopters could not be guaranteed, foresaw the termination of the survey if we did not disembark immediately. In a mad scramble, leaving behind most of what we had collected, we managed to board our boats and get away from the ship's side as she left the anchorage. One memory of the incident is of *Protector's* crew continually shutting watertight hatches on our heads, barring our exit, while we were opening them again to make our escape. We arrived back at camp after our replenishment visit carrying far less than that we had set out with.

In fairness to *Protector* and her officers she had a short season away from Portsmouth and after time spent on calls showing the flag in South America, and her time in the Falklands, her time spent south was so limited it could hardly be counted in weeks. With her officers being appointed for only two years, as was the standard naval practice, there was no time for any Antarctic experience to be gained. They were understandably, therefore, very cautious and very nervous. Their modus operandi in those waters was quite the opposite to ours. Whereas we took risks with the ship inasmuch as we accepted damage and were applauded for our struggles against nature in this unexplored region, the *Protector* officers feared enquiries, court-martials and almost certain ruination of their promotional chances were they to damage their ship. However even when they were standing offshore or at the edge of ice limits, their helicopters could be of invaluable assistance to the Survey, although they were limited to flying no further than 25 miles from the ship, and freezing conditions also curtailed their activities.

The Americans who visited came by helicopter from the coast guard icebreaker *Westwind*. They were wary of the approach to the anchorage, and did not seek what they later admitted to presuming to be our amateur help, but chose to remain offshore in the straits and fly their CO plus some others into us. Much to our amusement, those others included not just a foreign observer but also a rating to carry a coffee machine for his captain and a steward to dispense from it. They walked extensively over our island, taking much interest in our camp, and the surrounding wildlife, and were astonished at the professional survey we were conducting. We invited them to stay for lunch and with rather nervous glances at our cooking arrangements and the stew that was being prepared by one of the marines, they hesitantly agreed. Our mess tent sat all our party with room to spare, with the cooking apparatus sited at the far end from its entrance, and a central table down the remainder of its length, with planks supported alongside for seating. The American ship was, like all we met, dry. As we sat down to one of the Royals' variations of stew – with or without curry powder at one's choice – the Yankee CO surreptitiously produced from his hip pocket an illicit small flask of rum, which he proposed sharing with us all to mark their visit. Surprisingly we were able to produce some glasses and we all toasted the Queen and good old Uncle Sam. Top marks had to be given to our entire party for their diplomacy and composure until the Americans took off back to their enviable but 'dry' icebreaker, for during lunch they had been sitting on mess tent benches supported by five-gallon demi-johns of Pusser's rum!

The *Protector* did not leave the area but neither did she enter the Adelaide anchorage again. I temporarily left the survey party to pilot her across Marguerite Bay to visit Stonington, for her helicopter to assist with the work there. On passage they were again very nervous, particularly not liking my passing close by Guebriant Island, utilising a deep trench we had found rather than the inviting expanse of open water further off the island, which we suspected to be foul. On arrival off Millerand Island they balked at approaching Neny Fjord or proceeding to

Stonington Island for fear of the very strong winds that the area was notorious for. It was, therefore, a somewhat wasted outing and a slightly uncomfortable one too, for despite their friendliness, hospitality and obvious respect for my knowledge, I found the high number of nervous and noisy people on their bridge disconcerting, and likewise the agitated progress they made compared to the calm, almost silent, conduct of our small bridge team aboard *John Biscoe*.

In my absence the *Biscoe* had returned to the Adelaide anchorage and taking a new route from the eastward through a channel that is now named after me, had passed close south of Avian Island within close proximity of our camp. A grounded iceberg obstructed her entry to the anchorage, lying squarely in the narrow passage. Johnston, faced with choosing which side to pass it on, having to judge which side the berg had gone aground on, unusually for him chose the wrong side. He clipped a rock but despite heeling well over the ship suffered little damage. Had *Protector* remained in the anchorage I would have taken her out through that secondary eastern exit, with possibly disastrous results before Johnston 'found' that rock in what we had believed to be, although as yet unsurveyed, clear water.

Making use of *Protector*'s continued presence and the invaluable aid her helicopters could give to our survey we extended its limits by further triangulation. We visited Guebriant Island to take measurements, and were landed on its table-top-sized summit with part of our gear to await a second inward flight from the ship to deliver a tellurometer[44] and the heavy batteries that were required to operate it. Hovering over us to lower them, the helicopter lost its grip and plummeted down, striking the ground sufficiently hard to damage its undercarriage, which straddled Barry and me and two survey recorders. Seemingly worse was then, instead of lifting vertically off clear of us its pilot chose what was for him probably the easier and safer option of banking away into the air beside the small sharp summit, but this nearly dragged us off sideways with it. Incredibly, the helicopter's undercarriage only just touched us, and none of us were hurt.

The survey almost complete and the season coming to an end, we set about some last inter-lining and gap-filling, running sounding lines between those already made in areas where we wished to be sure there were no steep-to dangers between the existing lines, and to cover blank areas on our chart where icebergs had been, preventing our sounding the seabed beneath them. On the last day of our work, nearing the extremity of the survey ground before leaving the area, we ran a couple of lines where a berg had sat centrally in the main approach channel. The first line of soundings conformed to the seabed pattern and we turned to run the next parallel line. The bottom suddenly rose from fifty fathoms to five, and I thundered full astern and veered off to the adjacent line where there had been sufficient water. Johnston had gone below some time before, believing us to be well clear of all dangers. In ten fathoms, we lowered the workboat to make a search, finding a least depth of six feet. The berg aground there for most of our time in the area had been sitting against the sharpest of rock pinnacles right in the middle of the otherwise clear Johnston Passage. The danger is now on the chart as 'Full Astern Rock'. It provided an all-too-exciting climax to a couple of months of demanding but interesting work.

---

44 An electronic device to measure distance by means of microwaves.

The Argentine Islands

The low-lying and attractive Argentine Islands, divided by narrow boat channels and just beckoning to be explored, afforded us much pleasure in good weather. Climbing the gentle snow slopes of Galindez Island to its summit no more than 170 feet high for the outstanding views, we could ski in any off-duty time within sight of the ship, and explore the channels in our boats, landing and scrambling up the other islands, keeping us fit. It was sometimes easy to forget that we were, albeit in summer, in the Antarctic, and that both weather and ice could assault us here just as easily as elsewhere in the region.

The seasons of bad ice found us more than once either struggling to get to these islands and no further south, and that infrequently with the assistance of American icebreakers. They, liking the area, came south from Anvers Island base which they had taken over in 1964 under the control of their National Science Foundation. They also had a tendency to send their breakers round from the Ross Sea in years of difficult ice to practise their skills. On one occasion we had both the *Northwind* and the *Glacier*[45] leading us down the Penola Strait towards the islands. Following one of them extremely closely astern, as was the practice so as not to be left struggling through the debris of broken ice created but to go through it with them almost as one, we repeatedly collided with their fantail[46] when they were brought up all standing[47] by hitting particularly heavy ice. They were built for such nudging from astern and were comparatively unconcerned, but our 80 huskies, spanned out on deck and sleeping peacefully in the sunshine, would awake with a start at each collision, and would jump up and howl in chorus. We received humorous complaints about this from the American bridge.

They used their helicopters to plot their way through ice but not actually to con the ship from aloft. A story was told to us, when we paused and joined them by walking across the ice to each other, of how one American captain had handled his ship by giving rudder and engine movements to the officer on the bridge from the cockpit of the helicopter. Their breakers, with diesel-electric propulsion, had four propellers. Having ordered astern movements once or twice, he gave the order 'Half ahead starboard outer', to which the watch officer replied, 'I would sir, if we had one'! Thus the lesson to have one's feet on the deck to feel what is going on, which I emphatically endorse, as well as that of being able to go outside onto the wing of the bridge to hear, sense and feel the elements, and how they are affecting the ship and its handling.

The meetings on the ice always had a sub-plot besides exchanging information and pleasantries. Their ships were always devoid of liquor. To relieve that abstemious state of affairs on their part, when we came to rest at the end of a day's work close by each other in pack, a party would develop on the ice between the ships. The Americans were allowed to purchase crates of beer from us and drink them down on the ice. They would often light a fire, set out chairs and empty crates to sit on, bring hot dogs, and be joined by our crew to

---

45 US Coastguard and US Navy, respectively.

46 Overhanging deck, usually the main deck, at stern of ship (American).

47 From the days of sail: an unexpected stop so sudden that the sails were still set.

party. Their abstinence of some weeks was telling, particularly in comparison with our lads, they having kept in trim with their ration of two cans of beer a day. Neither did it remain the province of the ratings to become inebriated. I remember a very well-known and respected US Navy captain, a veteran of Operation Deepfreeze and later to become an admiral, visiting Johnston for such an evening. My cabin was sited at the foot of the companionway up to the master's flat. When the captain left Johnston at about midnight, he fell the full length of the companionway, ending up in my cabin doorway, where he slept till morning.

Their ships' performance was usually impressive, although it was interesting to note that at times in certain compact or unbroken ice they fared little better than we. First their weight bearing down on the ice would break it, and then their four propellers would churn that broken ice into brash. In very confined circumstances that would clog their progress, as does the deep natural brash which we called 'porridge', it enveloping and sticking to every inch along the waterline and sometimes to feet beneath; once, on their arrival to help us get through heavy fast ice, the voice over the radio advised, 'Stand back, *Johnnie Biscoe*, we're coming in'. Twenty-four hours later they were stuck about 100 yards ahead of us – and of course we had a party.

The Americans could not have been more helpful and pleasant to work with, but their coastguard cutters were far better and more professionally handled than their naval icebreaker, the *Glacier*. We also had much more in common with the coastguard crews, both of us being intensely interested in breaking new ground and surveying. I believe that their navy ship understandably suffered from the same problem as that of our own; short-term appointments, and therefore lack of experience down south. A further comment on the American operation was that they achieved only one quarter of that achieved by Britain in terms of published papers and reports, for the same expenditure. It was certainly evident to us, even on those occasional brief meetings, that they threw money at everything. One-second hand story of their operation on the opposite side of the continent was that aviation spirit was being transferred overnight from one of their ships to their Ross Sea base by pipeline for some considerable distance across fast ice. Come morning the officer of the watch taking over the bridge commented on the unusual pink of the ice between the ship and the shore. They had pumped umpteen thousands of gallons of pink-tinted aviation spirit onto the ice of the Ross Sea rather than into the station tanks. We were sometimes not so clever ourselves. The Argentine Islands belied its image of a polar paradise with the onset of a sudden prolonged and ferocious gale when we met both with the *Shackleton* and the *Kista Dan*, all squeezed into the far too small an anchorage off the base. The Lauritzen ship was carrying a de Havilland Beaver on deck with its wing tips protruding just beyond the ship's sides. With no room to manoeuvre, she dragged on to us and tore off an aircraft wing which then leached avgas that blew all over both ships.

It was during the stay of these three ships at this anchorage that a flight was taken in another aircraft by Fuchs and our second officer to observe the ice conditions south of where we lay. As they returned to the ships they were astounded to note that they saw the *Kista Dan*, painted red, some 30 miles before they could make out the *John Biscoe*, in the same anchorage, painted grey. It was decided there and then, despite Johnston's conservative wish to keep the *Biscoe* grey, to have our ships painted red.

Often the elements would seem to have one last onslaught against us at the very end of the season. At the Argentine Islands, anchored within the channel next to the base, we were gently harassed by small bergs, nothing too large being able to gain entry. But on the last dawn of the final call one season, we awoke to find a berg not molesting us, but sitting on our anchor. Grounded when under way on a spring high tide, it did not shift on the next high, and as the rise and fall of the tides reduced over the next few days we failed to free ourselves. Each day we shoved against it at high water, even the motor launch helping in its small way. We also tried driving our wash against it, with our bow against the shore. Explosives to cause it to calve and lighten sufficiently to float free might have been the answer, but we had none. With the season closing, we parted the anchor cable at a joining shackle, buoyed the short length of cable still attached to the anchor, and abandoned it for retrieval the following season. The seasons closed with remarkable regularity; for each base and locality we had a final date after which it was not sensible to remain for fear of becoming frozen in and having to overwinter. With temperatures dropping, if ice blew in and settled around the ship, particularly in tight anchorages such as amongst these islands, one never knew whether it would freeze in place and not shift again till spring, which, if caught within it, we would find rather embarrassing.

Such a winter freeze took place on a much larger scale. The Penola Strait, enclosed between islands, including between the Argentine Islands and the mainland, trapped a variety of ice that could become solidly frozen. On some years the Fids from the base where the vast majority of their work was carried out on site, rather than in the field, would travel across that ice to the mainland to explore the possibility of a route to the plateau, and even to venture further afield. Once during the period of 1959/64 the strait was crossed, the plateau reached and Mount Peary, 6,200 feet high, was climbed.

Shortly after we had sailed north minus our anchor to avoid entrapment, we received a message from the base that the berg had moved off our anchor, but also that Meek Channel, our anchorage, had frozen over. It transpired that it remained so for the winter. In later years, I used to keep a wary eye open at the end of the season on bergs going aground there, not so much on our anchor as in the two entrances to the channel, for had those been blocked there would have been no escape.

Deception Island and events of consequence afar

The difference in weather at the Argentine Islands, lying in the more usually stable cold air flowing off the mainland, and that of Deception Island, which lies in the path of the damp maritime westerlies, was very noticeable. At Deception Island most spare time was spent in hard labour. A hangar had been acquired in the late 1950s, intended for erection at Stonington Island in Marguerite Bay, to overwinter the aircraft. Two years of much ice had prevented its delivery there, and the decision was eventually made to erect it at Deception, from where the air operation was initially conducted. During the early sixties the work began, which first entailed digging through the permafrost to lay the foundations. I still suffer from unwisely volunteering to handle one of the pneumatic drills and consequently damaging my shoulder, probably making it worse by assisting in the erection of the hefty steel girders and iron cladding.

At about the same time, the other digging that took place, greatly beneficial to the ships, was that of a well. It had been suspected that water lay beneath the permafrost, kept fluid by the volcanic warmth. On going down 20 feet, water was found, and proved fresh although we aboard thought it slightly sulphurous. However, its discovery meant that this became the main watering place for the ships. We came fairly frequently to the island, due to it lying roughly on our usual north/south route and because of our involvement with the aircraft. During one such visit two significant events occurred in the world beyond ours; we usually paid scant attention to it, but these events were to greatly affect us all, particularly myself. FIDS became a public body, and the Antarctic Treaty was signed, thus securing our future.

Earlier that same season, whilst at Stanley, Johnston had given me the first of only two pieces of direct advice, as opposed to orders, that he ever proffered me. The Survey had evolved through Operation Tabarin to become a fragmented organisation under the Colonial Office, overseen in London by a combination of that office and our headquarters, and in Stanley and the south by the governor. A varying degree of interest and support was shown in our work and results by such bodies as the Foreign Office, the Royal Society and the Geographical Society, whereas the Colonial Office, which became integrated with the Foreign Office as the FCO, never fully understood our objectives and was rather uninterested. The responsibility for the scientific disciplines we were involved in lay either in universities or scientific organisations, such as Birmingham for geology and Edinburgh for geophysics, whilst glaciology had a unit within the Scott Polar Institute, and physiology in a division of the Medical Research Council. This resulted in our never being an entity with any satisfactory status, and no structure for permanent pensionable employment, or prospects for promotion for those ashore, let alone us seafarers. Then, worse, in a period of economy the Treasury proposed the closure of our entire organisation, causing Johnston to suggest that I might be wise to seek employment elsewhere. I did not take his advice, but stayed on in the hope of some better news. As mentioned above, that came on two fronts during a visit to Deception. Firstly, largely as a result of the highly successful IGY,[48] the Antarctic Treaty was ratified. Amongst others the Foreign Office and Royal Society then added their weight to the argument for our survival, which was agreed when put to the Cabinet. Secondly, we were made a public office and although a rarer breed of civil servants would be hard to find, it meant that both the science and our employment were put on a secure footing.

The Antarctic Treaty was signed in 1959 after some 60 meetings between the 12 nations who had operated stations there during the IGY. FIDS having maintained Halley Bay continuously since that year, the UK became the first signatory. The treaty came into force during 1961. One of the most successful of international treaties, it provides a legal framework for the management of Antarctica, and consequent meetings within that framework have resulted in agreement on the conservation of the continent's living resources and the protection of its environment. There are now 49 nations party to the treaty, it having 14 articles covering different aspects. The main articles are:

---

48 International Geophysical Year, 1957/58.

- It covers all land and ice shelves south of 60 degrees south latitude, but not the sea areas.

- Activities should be for peaceful purposes with no military activity other than to support science.

- Freedom and co-operation in scientific investigation.

- Free exchange of information and personnel.

- No territorial claims to be supported, enhanced or denied by activities in the area, and claims to be put in abeyance.

- No nuclear tests or disposal of nuclear waste.

- To safeguard the conduct of the treaty, observers from treaty nations to be entitled to inspect the installations and equipment of other treaty nations.

The treaty was established for 30 years and could, by agreement between parties, be adjusted, and renewed. The signatories included Argentina, Chile, Australia, New Zealand, the United Kingdom, the Soviet Union and the United States.

The Weddell Sea

There was always an air of excitement among those on board when leaving South Georgia for the several hundred-mile voyage, first across the Southern Ocean and then deep into the Weddell Sea, to reach Halley Bay Station on its eastern shore at 75° 31' S, 26° 37' W. Apprehension, too, of embarking upon a voyage encompassing a great degree of uncertainty, as we were attempting to venture to our normally farthest south in any season. Yet the base lies only 75 miles beyond that reached at sea in 1823 by the sealer Captain James Weddell, on a remarkably penetrative voyage in a year of little ice, nor too far beyond the place of besetment leading to the eventual crushing of Shackleton's *Endurance*. Opinions were exchanged as to where the northern limits of the pack ice might extend and questions asked of each other regarding its likely coverage and thickness, and whether any shore lead of open water to facilitate our passage would be found. These were all matters which would dictate the length of the voyage, the difficulties encountered and whether we might become damaged, possibly irreparably, meaning that the Fids would be unable to reach their winter quarters.

The Weddell Sea is a vast embayment of the southern continent, an extension of the Atlantic and Southern Oceans. Lying between Cape Norwegia and Joinville Island, some 1,000 miles apart, it is enclosed on three sides by the broad ice shelves bordering the peninsula to the west, those at its foot at approximately 80 degrees south, and the several hundred miles of ice cliff fringing the continent to the east. Without islands to disturb the pack ice, and a circulatory system and winds that can hold the pack ice within it for years, it is a formidable opponent, often requiring the ice to be worked into for more than 1,000 miles to reach our destination, making that a major achievement.

To make a passage successfully through the pack ice to Halley Bay, we had to have knowledge of the formation, lifecycle and movement of that pack, together with an understanding of the movement of the water currents that affected it, and the causes of the winds likely to bear upon it. To describe the passage planning through ice in this area, as opposed to the actual working of the ice in the immediate vicinity of the ship, also requires the interaction between currents, wind and ice to be explained. The circumpolar Southern Ocean current drives continually eastwards around the continent before the westerly wind down to somewhere in the region of 67 degrees south. Between there and the mainland at around 70 degrees south, a westerly flowing counter-current occurs, and as this counter-current arrives at the Weddell Sea and spreads into it around Cape Norwegia, it dips south and begins to form a clockwise circulation, the Weddell Sea Gyre. The pack ice forms as winter approaches, the sea first glazing over in hours of calm, and then quickly freezing to a depth of inches as the wind can no longer disturb the surface of the sea. It then grows slightly from above but more from beneath. Strong winds over any areas of water not completely frozen allow waves, and indeed swell, to get in amongst the new ice, pushing it around and rafting it, one piece upon another with a consequential increase in its depth, and leading to its often chaotic state. This is a simplified account of a continual process that creates moderately thick one-year-old ice. About 80 per cent of sea ice formed around the continent deteriorates and melts in summer, but in the Weddell Sea the very high percentage of winds from the north-east through east to south-east drive the southern Weddell Sea ice against the foot of the Antarctic Peninsula's eastern coast, causing it to remain in the area for years, and raft even further when under pressure. Multi-year ice of great depth, up to 40 feet thick, is thus formed. Some of this ice does eventually escape to the northward on the clockwise circulatory current, but much remains, providing a central core of very large, very heavy ten-tenths coverage, with relatively slightly lighter ice on its periphery.

It was as important to avoid the hard core of pack ice as to get as far south as possible before entering any concentrations. So a direct route from South Georgia could not be taken, but courses were laid off to pass just north of the South Sandwich Islands, thence south-east to about 60 degrees south, 20 west, hoping to skirt the northern solid edge of the pack ice, thence towards Cape Norwegia, where we might find less ice, or possibly even an open water shore lead running down the ice cliff coast, caused by offshore winds, enabling us to plan a passage towards Halley Bay. However 100 miles or so before reaching the coastal area, distinctly more eroded ice was found to lie running in a south-westerly direction, which could be successfully worked in a southerly direction; I attributed it to a warm, perhaps sub-surface, current associated with the commencement of the Weddell Sea Gyre. During some seasons of the late 1970s the ice deteriorated sufficiently for polynias to form. Later research found that the Maud Rise, an oceanic seamount lying in the path of the south-westerly current 200 hundred miles to the north, disturbed its flow. This created an upwelling of warm water through the cold top layer, causing the ice deterioration that we sometimes found. When this phenomenon did not occur we made for a possible shore lead. The prevailing winds south of the main depressions, and the flow of cold air down and off the continental ice dome are both of an easterly direction causing ice to move away from the coast and create leads through the pack or even completely open water.

In my early days we knew very little of all this, but gradually over the seasons we learnt to understand what was going on, but often not why. In practice, too, the variations in the disposition of sea ice were large. The shore lead can fluctuate within a season from yards in width to more than one mile, some hair-raising passages being made to stay in open water when a recently calved piece of shelf ice lay only yards off its parent cliff as a large tabular berg, providing a gap to pass through, whilst solid pack lay against the seaward side of the new berg. In some seasons we were confronted with heavy pack for our entire passage. We once took six weeks to make good the 1,500 miles to Halley from South Georgia, but another time I took only six days. In the first years of making such passages there was a lot of trial and error, much exploratory steaming, and a great deal of damage sustained. The *John Biscoe* had to be repaired and considerably strengthened after her first voyage into this region. The circuitous route towards Cape Norwegia became established practice fairly early in the relief years of Halley Bay, but apart from knowing that the offshore winds created shore leads we knew little of the mechanics of the ice distribution.

The immediate working of pack ice was basic; espying and making towards water sky, finding leads to run down, and finding ice to negotiate in places where, wind-driven past icebergs and fragmented, it was more workable. Patiently forcing and attempting to widen existing cracks although mostly done at full power was, without much impact, less damaging than ice-breaking. Going astern to create room to charge and hopefully crack floes was the last resort, putting the stern gear at risk and with the likelihood of damage to the forepart. More sophistication came in later years with satellite imagery of the ice cover, and a stronger and more powerful ship.

The navigation in pack ice during my years was also fundamental, never having satellites to utilise, but in retrospect it was all the more satisfying. With the sun and planets providing sextant sights for celestial navigation, but often obscured by cloud and frequently without clear horizons due to ice of any nature obscuring them and perpetual daylight prohibiting the use of stars, dead reckoning[49] was our prime method of navigating, and we became very adept at it. Whereas the normal estimations of set and drift could be made in the open sea, an understanding of the movement of pack ice and the importance of its drift was required, which always bore a ship off course to a greater degree than expected, exacerbated when working through it at slow speed. Pack ice driven before the wind often attained speeds up to 6 knots and occasionally more, whilst the gale influencing it might be many hundreds of miles distant, and not evident to us. Because of the Coriolis effect[50] the pack would not drive straight downwind, but would make good a course, and thus our drift within it, of about 30 degrees to the left of the wind direction. If one was beset during a gale, such a course and speed was made good whether one wished it or not. The utilisation of large icebergs, nearly always present, to base a dead reckoning on was the solution to an otherwise difficult estimation of

---

49 The method by which a ship's position is derived from the ship's courses and speeds plus the estimation of the wind and current affecting that made good, but without the benefit of astro fixes.

50 The earth's rotation to the east deflects anything moving freely above or on the surface of the earth to the right of its original direction in the nirthern hemisphere, and to the left in the southern, the effect being greatest towards the poles.

progress and position; because their vast underwater bulk presented such an impediment to their movement that they were virtually stationary. Continually plotting the position of the vessel in relation to a chosen berg on a blank sheet, then transferring a day's work onto the chart, proved very reliable.

One technique that could never be improved by any later sophistication of either ship or navigational equipment was that of entering a field of ice from seaward; that final commitment to the pack ice when no progress can be made towards the destination without entering it. Much patient steaming along the ice edge, more usually a windward one, perhaps in a rough sea with a high degree of movement, is required. The pack beyond the edge, with any sea or swell visibly dampened, can be an inviting prospect, but entry at a protrusion of its edge could waste hours unnecessarily working miles of ice, together with the use of precious fuel. A pattern of embayments in the ice edge is usually formed, and the deepest found and utilised for entry, but that is only the beginning of the exercise. Pack ice blown downwind has its lighter pieces driven ahead, they offering less resistance to water, meaning that the heavier pieces lie to windward, the heaviest being on the windward extremity, exactly where it is normal to gain entry. Any swell running into the ice edge rebounds, as off a cliff face, causing a confused and lumpy sea through which the ice has to be entered. Good speed is needed to get through the initially heaving, heavy outer pieces of ice as fast as possible, yet in order to do so without damage, caution is required not to strike the lurching lumps too hard. Whether to strike a single relatively lighter piece, if at all possible, to crack and divide it and thus to progress, or to choose a gap between floes, risking striking them on the entrance[51] – normally bad practice – is another decision to be made. I do not believe there can be any hard and fast rules for such a manoeuvre, but there is usually a great deal of apprehension as the ship drives into a wildly seething mass of heavy ice, followed by a great feeling of tranquillity when out of the swell, and a sense of achievement at having entered an ice-field without damage.

Halley Bay

Halley Bay Station and the embayment in the ice cliff coastline beyond which it lies were named after the great mathematician and Astronomer Royal, Sir Edmund Halley; it is situated on the Brunt Ice Shelf, which forms part of the eastern shores of the Weddell Sea. It was established by the *Tottan*, a small Norwegian sealer, on behalf of the Royal Society, as part of the International Geophysical Year Programme of 1956. On completion of that programme in 1959, the station was handed over to FIDS, and a Dan vessel of Lauritzen Lines was chartered annually for its relief and resupply. When the *Shackleton* was acquired in 1955, she had enough cargo space and accommodation to carry out this task, but was never intended for that role, being neither strong nor powerful enough to work the pack ice of the Weddell Sea. The new *John Biscoe*, however, could do so as long as there was not to be a rebuild of the station. When there was, a Dan charter vessel would sail in tandem with the *Biscoe*, and together they would carry out the rebuild and relief. Due to the nature of the site a rebuild would recur every several years.

---

51 Where the arrow-head form of the bow becomes the full breadth of the vessel.

The ice cap or dome of Lesser Antarctica, centrally some 10,000 feet deep, flows at about 700 kilometres per year from its lofty interior towards the coast, where it mostly terminates in ice cliffs of between 50 and 200 feet in height. It is mainly supported by the fragmented rock base of the Antarctic continent, which in places also rises through and above the ice cap to form the Trans-Antarctic and Ellsworth Mountains, and many nunataks,[52] around which the ice flows. Powered seaward by its own weight whilst reducing in thickness, the ice cap sometimes falls gently as snow slopes, blasted by wind into undulating sastrugi,[53] or more steeply as ice falls or glaciers, distorted and crevassed, as dictated by the terrain beneath. It reaches a point where it is no longer supported beneath at the so-called hinge zone. Here, now relatively shallow as the rock base beneath falls away into deep water, the shelf becomes suspended in the sea, partly held afloat and partly held in tension with its parent ice shelf. When movement from seaward disturbs those conditions or the elasticity of the whole mass fails, calving takes place and icebergs are formed. These vary in size from small fragments to tabular icebergs of up to many miles in breadth and width with heights exceeding 200 feet above sea level and five to six times that beneath.

That calving of the shelf is the first reason why the base has to be rebuilt every several years. At the time of my visits the site lay about two miles from the shelf edge to reduce the risk of it being within any calving area and thus being launched out to sea, yet no further inland, as this would have compromised the satisfactory relief of the station from the ship. The second reason for the rebuilds was that snow drifted and accumulated above any stationary object, be it a box of stores, an oil drum or huts, upon the vast ice shelf. In addition, the natural annual accumulation of snow over the marginal edges of the dome was about four to five feet a year, burying a base to a dangerous degree; the accumulation of snow turning to ice above a base tended to crush it, and also required the entry to the base to be kept open through a deep vertical access shaft down through the ice, creating a very unsatisfactory situation in the event of fire, a dreaded occurrence at any Antarctic station. In addition, the warmth of the whole complex would create excessive water, which would drip from the roof of the virtual ice cave entombing the base, and this would contribute to making the huts untenable after some years.

The general location of the station was chosen for its suitability for geophysical and atmospheric investigations, as was the expense of repeated rebuilding and re-supplying it justified by the location being of prime importance in the aural zone. Glaciology, meteorology, and geology by field parties ranging from the base, were undertaken there. The precise positioning of the 20 to 30-man complex was initially determined by its access from seaward, and remained so subsequently, during my time of some further rebuilds.

A feature of the Brunt Ice Shelf, created as it is by the continental mass of ice pushing it afloat out to sea, is that it grounds again on a rise of the bedrock, producing fractures and crevasses in an area known as the McDonald Ice Rumples. This results in coastal fissures forming in the fashion of spreading fingers with elongated creeks between them, as the ice rises over and around the rock beneath. In winter these become frozen over with fast ice. The

---

52 Small isolated peaks of rock protruding through an ice field or glacier.

53 Wave-like configuration on the surface of hard snow formed by wind.

prevailing winds, enhanced by the flow of cold air falling off the continent, cause snow to drift off the ice shelf and fall down onto the fast ice below, forming a natural ramp between the two. In January or February, when the ship would arrive, the expectation was that although the pack ice out to sea was beginning to fragment, there would be fast ice with a fairly flat surface still attached to the cliffs, and a ramp from it to the cliff top, remaining within the fingers. The low fast ice would afford us a place to berth at, and the natural ramp would provide a route to take our cargo up on the last leg of its journey to the base site. Occasionally impenetrable fast ice remained out to sea causing us to discharge some miles from the coast, hugely increasing the journey the tractors had to make to the base. Conversely the swell sometimes removed both fast ice and ramp, and we were forced to lie against the ice cliff to discharge cargo onto the top of the shelf above bridge level.

Emperor penguins

Of the two dozen sites on the Antarctic coasts where emperor penguin colonies are to be found, Halley Bay was one where they formed regularly on the sea ice at the foot of the ice cliffs. On our visits there during January and February we saw few, but during our approach sometimes many, with their downy chicks travelling north over the disintegrating pack ice to feed on the summer plankton as they found increasing amounts of water between the floes until they reached the open sea. Their extraordinary breeding cycle accounted for these sightings offshore and the penguins' absence from the rookery on our arrival. For the chicks not to have to endure the harsh midwinter, but to have a chance to mature in the most clement weather, the emperor lays its eggs in autumn. The male birds then incubate them for about two months during the winter, each resting an egg on its feet to keep it off the ice, enveloping it within a fold of feathered abdominal skin. Meanwhile, the females go off across the pack in search of food, returning with it as the chicks are about to hatch. The females takes over the care and feeding of the newly hatched chicks from the part-digested food in their crop, whilst the males which have starved for the two-month period of incubating the eggs plus a long period during courtship prior to that, go to sea to feed for three to four weeks. They then return to feed the chick, and the mothers depart to feed. By spring the cycle is greatly speeded up such that one parent cares for the chick alternately for about three to four days whilst the other goes off to feed. Eventually the chicks, having almost reached independence, depart seaward together with the parent birds in spring, to be observed by us. Their habit of incubating the eggs on sea ice is probably because the sea beneath it makes it marginally warmer than the shelf ice ashore.

One year our only sighting of these amazing birds was two abandoned waifs, small and very frail, and still totally covered in their thick woolly down, showing no signs of their ultimate height or the splendid adult plumage of brilliant orange ear patches and light yellow shirt fronts beneath a black cap and blue/grey back. As they had quite obviously been left to die, we took them aboard and with difficulty fed them the nearest concoction we could produce to their naturally acquired plankton. They flourished so well aboard, and we became so attached to them, that we were loath to place them in the care of a base as originally intended, but kept them on deck until we reached Montevideo. From there, Pinky and Perky were flown to a northern hemisphere zoo, where I understood they continued to thrive.

One of the most remarkable sightings of these wonderful creatures occurred just after we had departed north from Halley Bay a little later in the year than normal. Over the horizon across the ice two widely separated but converging lines of penguins approached us, each bird diligently following the other, waddling or temporarily tobogganing on their stomachs. That year the rookery had been in two distinct sections separated by a few hundred yards. Each line of emperors must have contained birds from only one section of the rookery, but for each line of penguins to regain the place where they had emerged from the sea, they had to cross the other line. They passed between each other almost perfectly in turn without a single one ever leaving its own line to join the other. They would have been returning to their section of the rookery to find their partner, more often than not that of the previous season, commence their courting and start the whole incredible life cycle again, with the laying of eggs some months later.

8 Command of the
RRS *John Biscoe*, 1964-1969

First voyage

In November 1964, as we embarked upon a further voyage south, Johnston stubbed his toe when taking a shower. Between Montevideo and Stanley it became badly swollen, and he requested I dress it for him, not wishing his plight to be generally known despite there being a doctor aboard. Arriving at Stanley his foot appeared to be infected. He had a history of problems with circulation in his legs and feet having, I believe, being slightly wounded in them by shrapnel during his war service. I nevertheless, perhaps naively, or because I could not quite visualise the *Biscoe* without him, believed that at the worst he might spend a few days in the Stanley Hospital, and we would have our stay in port extended whilst he recovered. The senior medical officer saw him as soon as we berthed on the morning tide, and I was advised that gangrene had set in, and that he considered that Johnston should return to the UK for treatment. There was no air connection to the outside world at that time, just the mail steamer to Montevideo, the RMS *Darwin*, and as she was due to sail on the very same tide we had berthed on that morning, Johnston was advised to take that opportunity to leave. I went to his cabin and whilst his steward packed a small case with what he might require for the journey, he ran through the legal and financial necessities, the former overseen by the shipping master. I then took over the ship's papers, accounts and official logbook.

Johnston had always been a strict disciplinarian, applying his severe strictures of behaviour and dress upon himself as well as his officers and crew. Aboard he always wore uniform, even in the severest weather, and never did rounds of the ship or inspect the bases without wearing his cap. To say goodbye to him at the foot of the gangway in a wheelchair, in his brown dressing gown, one foot bandaged, the other in a carpet slipper, knowing how humiliating it must have been for him, was sad. As we finally shook hands, he gave me one piece of advice, only the second in all my time with him: 'Don't change anything for a season.'

This advice – unlike his first, advising me to think of seeking employment elsewhere – I took, and it served me well. I sailed south two days later as master, temporarily short of one officer. The second officer, now acting as first, who went forward to raise the anchor, did not time the heaving by the windlass correctly to assist the ship coming off the berth, but began to drag the anchor home and needed guidance. I immediately became aware of how little experience he had, and that this was to continue throughout that voyage; a sure reason to heed Johnston's advice and leave his systems and routines alone, for I would have enough to

128

think about without trying to set up any of my own. During my five years as chief officer I had often thought I knew better than him, and when I took command, I would make changes. But when the time came to have the responsibility of command rather than that of number two, I soon came to realise, as I am sure many before me have found, that from this new viewpoint matters were seen very differently and were indeed best left alone at first. So at that point I ran my ship his way.

I was advised of his deteriorating health and ultimate death in hospital in London during the voyage. Understandably no comment was made regarding my permanency as master, so this voyage would have seemed at first rather as if I was on trial, but such was the utterly absorbing nature of the job that it never did so. Neither did it concern me or affect any of my actions or decisions. To my fellow officers I had always been Tom, to the crew the mate. They all now compromised and I became known as Captain Tom – amongst almost certainly much less flattering titles. Mostly all were very supportive. The chief engineer, Digger Ward, had been alone in being of a similar age to Bill Johnston and missed his former captain dreadfully. He appeared lost when it came to gin and cribbage time, and became somewhat of an isolated figure. I could never have provided him with the companionship that Johnston had. At first he would grumble at my initial slowness or cautiousness, but never once questioned my decisions or requirements, and soon became a very loyal supporter, continuing to run his beloved engines without a moment's failure whilst maintaining his extraordinary record of starting them in an emergency in less than a minute. One young deck officer of just a single season's experience did become very surly and awkward, resenting my new authority; I have often wondered if he had ever understood the weight of responsibility on my shoulders, our being shorthanded and he being the only deck officer other than myself with any southern experience. Nevertheless the first part of the voyage went well; just enough ice, poor weather and difficult anchorages to test but not overtax me – though tax, in the other sense of the word, did concern me, for one of the jobs I had inherited was to compile the crew's wages. We paid Falkland Island tax directly in cash to the Treasury, on changing Articles at the end of each voyage, so there could be no delaying the bookwork or indeed no time to seek any advice. All had to be squared off by the time we reached Stanley for the final visit. Laying out the various tables that had to be consulted to find the deductions to be made to each individual's wages, left too late as we heaved our way across the Drake Passage, provided me with some of my less placid moments.

Another first for me was the utilisation of the Code Books, which was none too difficult being one-use pads, but very time-consuming. A message headed 'For your eyes only', usually from the governor, who seemed to relish a lengthy use of the code on the most trivial of matters, would arrive in the middle of some other activity that needed my heightened attention. I enjoyed the total responsibility and very much warmed to my role as father figure. I began to learn the philosophy of being a good shipmaster on long, isolated voyages and to be firm yet approachable. Some aboard were older than me, and some mess and deck boys barely 16. I was surprised, whether they were older or younger, at the number of financial, marital and other problems they brought to me. I was also astonished at the trivia with which wives would worry their husbands. I would nearly always know of an impending problem because I was assisted by the radio officer who saw every message sent or received. As well as acting as my secretary and knowing all my business, he would also have a quiet word with me if he foresaw a problem arising elsewhere. He was a paragon of discretion, always telling me what I ought to know, yet sheltering me from nonsense when I was busy. Because of the lack of experience aboard, I soon became a harsh disciplinarian; my way was the only way, for only I had the relative experience of the most demanding conditions and circumstances that were met. Both at sea and lying at the bases I maintained a rigorous regard to safety. I allowed no loose interpretation of my instructions or any departure from them, which made me a tyrant in the eyes of some. Thus my style of command was formed not out of arrogance but of necessity, but remained with me until I retired.

When we returned to Port Stanley around mid-season we were met by the *Kista Dan* on her way from Montevideo to Halley Bay. Aboard was Malcolm Phelps, my replacement as chief officer, whom I was very pleased to see. A few years older than me, he held a master's certificate and had been recruited in London, then flown out to join the *Kista Dan* in Montevideo. Malcolm was solid and dependable, and little got him excited or unnerved. At times his quiet, dogged, almost slow approach to work could be frustrating, almost infuriating. Then I would reflect on how sound he was, and let him get on with things at his own pace. He was probably the perfect antithesis for myself who, as I well know, am inclined to be impatient with people, though not circumstances. Malcolm later became a very successful and respected master of the *John Biscoe,* and served for about 20 years with the Survey until retirement.

Surveying off the South Orkney Islands

We sailed after Christmas from Stanley with Malcolm aboard as mate. Our destination was the South Orkney Islands; we were to survey the coastal waters immediately to their south. We had aboard a naval hydrographic survey party of seven men, led by Lieutenant Commander Joe Bradley, and a number of Decca specialists who together would conduct the survey under my command. This called for close co-operation between all aboard. Joe and I got on particularly well, although naval hydrographic surveyors are a particularly dedicated and an immensely keen breed, always going – and wanting – that extra mile, sometimes literally closer to the shore, for on this occasion all sounding work was initially to be done from the ship.

The survey was to evaluate an electronic position fixing system, Decca Hifix, on one of its first field trials. This consisted of a master receiving station situated in the chart room abaft the bridge, with an aerial erected at the after end of the ship. Many miles apart ashore we set up three transmitting stations under canvas with their own generators, to cover our intended survey area. A Hifix technician and two assistants manned each land station. On board we had the expert who kept the whole system running, but there were many technical hitches that involved a great deal of revisiting the sites in the inflatable boats. The transmitters put out concentric circles and the on-board receiver displayed which circles we were situated on at any one time. We noted the readings and from them plotted our position on a plan of these circles (lanes) that we had drawn. Results with an accuracy of within two metres were obtained. Conning the ship to conduct the survey, we were able to lock onto a lane from one station, the helmsman then steering round that lane by watching a dial which indicated to him whether to go to port or starboard in order to maintain the lane adopted. So sensitive were the instruments that when setting up the system it confused us all, for when we altered course to follow the needle's indication to regain the line off which we had wandered, the dials of the master station moved as if we were altering away from it. This was because when the bridge went to starboard, the stern, where the aerial was situated, went to port at first, the ship pivoting about its longitudinal centre. When the difficulties of keeping the stations on the air were overcome, we progressed the sounding with a speed, assuredness and accuracy never remotely achieved before in these waters. Additionally we were able to survey during darkness, and in snow and fog; splendid for our productivity, but wearing for me. Surprisingly, one unforeseen problem was grounded icebergs that blanked the radio signals from the out stations. The area being notorious for the abundance of bergs emanating from the Weddell Sea, we assumed that the manufacturers would either have taken this into account, or would not have had us evaluate the system in this area. The bergs continually moved on high tides, and having firstly prevented us sounding where they sat, they then shifted to allow us to sound that area, only to ground again and deny us access to another. The result was that our fair chart looked as if we were tackling the task in a most disorganised fashion, and brought about considerable delays. The disruptions and changes of plan caused by this and the downtime of the transmitting stations made Joe an extremely frustrated man, fearing that the new chart would never be completed.

There was one tabular berg, about six miles in length and two in breadth, that remained firmly aground on our site for nearly the whole period of the survey. Its removal came about with some drama. The *Shackleton*, under the command of David Turnbull, was visiting the nearby Signy base when a very strong gale blew up, and the option arose for both of us to utilise the shelter afforded by the iceberg. It could be extraordinarily quiet out of the wind, sea and swell, compared to that of the nearby exposed water, although it was not always a simple task to remain there. The usual methods were either to steam broadside onto the wind under the lee of a larger berg, or to steam upwind towards a smaller one, before dropping back to repeat the process. But whichever method was utilised, it entailed constant manoeuvring and frequently being extremely close to a berg to be effective. Not surprisingly, both David and I chose to tuck ourselves under the lee of the six-mile berg. Shortly after taking shelter, the placid mammoth of a berg, immune until then from wind or tide, suddenly fragmented, to replace itself with a cauldron of seething ice. Tabular bergs a half-mile in breadth formed, maintaining the same stable attitude as their parent, but separating from it at speeds through the water never previously witnessed, driving a tidal wave before them. Smaller bergs calved more chaotically, collapsed into the sea, gyrating madly as they sought to find equilibrium. Blocks of ice the size of houses, freed from below, became buoyant and surged to the surface rolling violently, whilst the sea all around us became filled with bergy bits and brash ice moving erratically in the maelstrom created by the major disturbances. The *Shackleton*, one moment within our view, was nowhere to be seen.

On a few occasions in my career I felt despair; this was one of them. On my bridge we all thought that the *Shackleton* had gone down, slammed into by the calving ice, or driven against it, or overwhelmed by the enormous surges that had been created. We called on the radio but received no reply, and began to steam amongst the still-shifting ice to look for her or her wreckage and survivors. Then after seemingly a long while she reappeared, as pretty as ever, apparently completely unscathed. The numerous new smaller tabular bergs had prevented any radio contact with her, which we had completely overlooked in our concern.

The break-up of the giant berg did allow us to complete most of our survey, although to this day large areas within pecked lines can be seen on the Admiralty chart where bergs impeded our sounding. Shortly after this incident we returned to Signy base for the final visit of the season.

Caught within too tight an anchorage

A customary exercise on the last visit of a season was, if possible, to top up the fuel of a base. At Signy, unless lying against fast ice, we normally lay to one anchor in the centre of Borge Bay. The base had a good tank near the shore, so to discharge fuel I went in deep, across the wind, just short of Billie Rocks on its northern extremity, dropping two anchors well apart, one after the other. Then we paid out eight shackles on each, the maximum scope the width of the bay would allow, dropping astern to the leeward shoreline, into Factory Cove where the tank was sited. Mooring lines were then run ashore, so we ended up lying tight between a pair of anchors and lines astern with the rocky shoreline barely 30 feet away. The refuelling hose was run out through hoops along one of the mooring lines and pumping commenced. We had sighted heavy pack ice to the south of the islands and I was keen not to delay, for that pack, coming as it was from the Weddell Sea, would be heavy and could present us with a major problem were it to surround the islands. The date was the one fixed in my mind from my experience of when the ice might begin to consolidate and remain for the winter. It was also to be my last task of the season and we were due to leave the Antarctic as soon as it was completed.

During the afternoon weather reports from the bases on the peninsula, to our westward, indicated that a severe depression was forming to their west. I took note, but judged that it would pass so that the gale would blow from the direction of the base and the tank that we were refuelling. Also, because the island terrain there would give us some protection, I decided to remain where I was, but I backed up the mooring lines with our best wire hawser to hang on with. I intended to finish the job and leave.

But the depression deepened and did not track as I had anticipated. The wind came from the opposite direction to that expected, from where our anchors lay, pulling them down the sloping seabed from Billie Rocks towards the deeper centre of Factory Cove. Much more severe than expected, the wind soon made the anchors drag appreciably, and as darkness fell with so little clearance astern, I decided to get out into the open sea. But by the time we had uncoupled the hose and let go our lines, the stern was only feet from the rocks. I worked the engines and helm constantly to assist Malcolm on the forecastle to lift the cables, but repeatedly, when they became vertical with no strain upon them so he could take in some cable, we fell off the wind, dragging shoreward. Again and again, over some hours, I desperately attempted to lift the anchors yet stay off the shore.

Evening turned into night with the anemometer needle often hitting the 120-knot stop, and after an exhausting number of manoeuvres, both anchors were finally weighed. Then the most dangerous part of the operation began. The entrance to the bay was only 700 feet in width between the visible dangers of Small Rock, one foot above water to the north, and Bare Rock, nine feet in height to the south, but with less than 200 feet of shoal-free water to pass through. I had the basic outline of the cove and bay on my radar, and knew by heart the courses to be kept and the various alignments of features to assist, but it was pitch dark and the sea was strewn with bergy bits, some under way in the storm and others aground, cluttering the radar screen. Bare Rock was indistinguishable from them, and Small Rock, awash in the sea that was running directly across the entrance, totally indistinguishable. I could not proceed out too slowly for fear of drifting downwind onto Bare Rock, yet at times was pointing as much as 30 degrees up into the wind. With one officer on the echo sounder screaming out the depth to me above the noise of the wind, I ran repeatedly from the radar to each wing of the bridge to look for those gateway hazards, basing my actions on the lie of the land as shown by the radar astern and my mental picture of the usually attractive anchorage, but seeing neither entrance rock.

But then the depth began to increase, as did our radar distance from the coast, and I knew we were clear. I took the ship about five miles off the coast, hove to and awaited the storm to ease. I went below at 0400 hours and caught some rest, having decided to get out of the anchorage some 12 hours earlier. The next day the wind dropped almost completely, and being northerly had kept the pack ice offshore. The sun shone and it was hard to believe the conditions of the night before and the danger we had been in. For a day I basked in the glory of a job well done. The noise of the wind and the jarring of the anchor cable, and the incredible number of engine movements had kept all below well aware of our predicament. Joe Bradley had observed the early hours of the incident from the after end of the bridge before going below, and was most gracious with his compliments. Then I began to wonder, then doubt severely, my own actions. With hindsight I had made the wrong decision; I had taken a terrible risk in leaving the cove past those rocks. Had I holed the ship as we passed between those unseen dangers, with no room for the slightest error, we could have been in desperate trouble as we entered the deeper water beyond. I should have eased the ship ashore as best I could within the anchorage, accepting the likely damage, leaving the anchors in place to haul off later. We would have been aground, with a few holes – but alive and ready to tackle another day!

We returned to Stanley for some final fun and games, I had had an exciting season, demanding but apart from the incident at Signy Island, not worrying. My spirits high, by day I enjoyed shooting hares on Mount Tumbledown, collecting mussels from the hulk of the Great Britain and walking the lovely beach at Gypsy Cove, and in the evening going to parties, to happily conclude the season. I also had cause to visit the venerable Scottish medical officer, who had formerly been a Fid, to treat some boils around my collar line. Whilst he was lancing them he said to me 'Dinnae worry yourself, laddie, it's a common manifestation of stress,' and my ego as a cool, young captain was rightly deflated!

Successive voyages

My next voyages in command continued in the usual pattern, visiting both the peninsula and Halley Bay, sometimes taking the role of relieving Halley on our own and sometimes in the company of the *Perla Dan*. Sailing from South Georgia only a day apart, we each evolved our own strategy as to which route to best navigate the ice of the Weddell Sea. There was no question of one vessel closely following the other in the style of an icebreaker working in tandem with a cargo ship, for neither of us was powerful enough to clear the way for the other. Sometimes we were on the horizon to each other, sometimes within walking distance over the pack. It became customary on these latter occasions for their master or me to walk over to the other's ship at lunch time, if the ice was particularly tight and little progress being made, to discuss the conditions and have a drink together. My offering was a glass of beer or a gin, but his was a tray of several types of whisky, sweet and dry vermouth and a jar of maraschino cherries, and of course ice. From this I helped myself to the quantity, strength and taste of my liking, and in accordance with the afternoon ahead. I habitually drink that variation of a Manhattan to this day.

The other important discovery of that voyage was that as we progressed, leads and water sky led me towards the central part of the Weddell Sea, where the core of heavy, almost static multi-year ice usually lay. This was what I had been drawn to, and so reprimanded for following, on that earlier voyage as chief officer. This time it was upon my own head, but led to the discovery of a polynia which gave me a 200-mile almost open-water run where solid ice was always thought to have been, and then into broken ice all the way to Halley Bay. Crossing a polynia might often be done by keeping fairly closely to the side of the open water that lay towards one's destination. This sometimes provided one of the great natural sights at sea in these southern latitudes. Krill, already in superabundance in this part of the Weddell Sea, move with the surface water and become caught against the pack ice edge in vast quantities, often with the spread of football pitches and solidly to a depth of six feet. Baleen whales, including the huge blue whale plus the humpback, the fin, the sei, the southern right whale and the minke, the latter being the most common, could be seen by the score and sometimes the hundreds, gently undulating through the swarm of krill, sucking in mouthfuls as they went. The decline of all whales throughout the sixties from overfishing was noticeable and depressing, such that in my latter voyages it became a special event to see them at all.

During a later season, on going to Halley Bay with the *Perla Dan* finding a polynia was realised to be not such a rare occurrence. This time, however, a polynia was spotted, reached and utilised for the first time by the use of satellite photography. We had left South Georgia four days astern of the *Perla Dan*, which took the customary eastwards route towards the coast despite our both being advised by Fuchs in London, from viewing satellite photographs obtained from America, that he thought that he could discern through the cloud cover an expanse of open water some two to three hundred miles further to the south and west of the two ships. I followed his advice, overtook the *Perla Dan*, which became beset on the more 'normal' route, and again having found a vast polynia, arrived four days ahead of her. My message to HQ was that it was like having 'Old Father Fuchs up there in the sky, smoking his pipe and looking down on our every move'. A new age had dawned in the navigation of vast areas of ice, particularly the Weddell Sea, whereby it was possible to plan the strategy of such a voyage before entering the ice at all, and then with updates, select the general direction of where best to head.

The publicity machine of the Survey was almost non-existent, but occasionally a bulletin was given out to make the public aware of our activities and achievements, to balance the inevitable reports when accidents occurred or things went wrong. We had no knowledge of a press release being issued regarding us when beset for two weeks in the Weddell Sea pack, until we heard it on the radio. A severe gale had blown up whilst we were already in very heavy ice. The wind screeched and howled through the masts, rigging and aerials as we lay virtually motionless within the ice and, with no main engines running, almost silent. The snow lifted from the pack, and driven as a blizzard, obscured everything but the immediate ship from view, with one exception; some emperor penguins, probably finding some warmth and shelter from our presence, collected close under our lee side, having sought us out in their usual inquisitive way. The pressure built up considerably, and ice blocks were both rafted alongside, and driven beneath us. The consequent movement of the pack disturbed our new neighbours, who perpetually shuffled to resettle at each upheaval, apparently annoyed at having to do so.

There was little we could do at the height of the storm except hope that our ship was strong enough, which it had mostly been since some extensive re-strengthening after its first voyages. Intermediate frames had been introduced between each of the main frames together with the occasional deep web frame and some longitudinal stringers connected to the web frames by deep knees. Our besetment lasted over a weekend so there was no reason not to indulge in our usual Saturday night dinner and film show in the wardroom, which was our practice when there were no other demands on us. After dinner we 'passed the port' and with our entire number present, some smoking (I a small cigar), we opened the leeward portholes, for it was getting too hot and airless. As 2000 hours approached the radio officer put the BBC news over the speaker system whilst the electrician set up the film projector. Suddenly we were aware that there was a report that the RRS *John Biscoe* was in dire straits, bravely battling hurricane-force winds amidst greatly pressurised ice in the deepest south! We drank to her release and enjoyed the film, and fortunately suffered no damage on that occasion.

Measurement and effect of swell

On my early voyages into the Weddell Sea, some studies were made of sea ice. Associated with this was the development of a wave height recording apparatus, which promised to be useful for navigation in ice. This device, fitted to the hull, could sense the minute fluctuations of the swell when deep in pack ice. It indicated the presence of open water at some great distance well beyond the horizon – but how far away, and at what height the swell was on entering the pack was not shown, so although it was of interest it was ultimately of no use to us. The direction of any swell was easily discernable by eye when it was just a few inches in height; it was fascinating to see a whole field of ice gently undulating, and this could be of use when confronted with alternatives as to which direction to head. The device was modified later to measure open ocean swell, recording heights of over 70 feet.

In January 1967 we visited Halley Bay in company with the *Perla Dan*, when the base somewhat dramatically, had cause to change its name. On our arrival, the swell, in a season of light ice coverage, had already removed much of the shelf ice in the vicinity of the base, including the headlands forming the bay, transforming the immediate coastline. The fast ice and the ramp within it were naturally also gone, denying us our usual access. The two ships had no alternative but to secure alongside the ice cliff to discharge the materials for a rebuild of the base. The cliff, further disturbed by the swell repeatedly calved, forcing us to leave our berth and re-moor. It was interesting for me to have to make way as quickly as possible through massive blocks of shelf ice to seek yet another cliff face to lie against without nudging and unsettling it. Of concern, though, for those waiting to assist us in mooring as we manoeuvred into position, was that the cliff would calve again – which it did, more than once. Our method of securing the ship to the 'shore' was to run out mooring lines which were then attached to baulks of timber set at right angles to the line, and placed in holes about six feet deep in the ice. These were refilled with snow and trampled upon; then the final technical and necessary part of creating those deadmen was for the shore party to urinate over them, providing instant cementing. Those placing one deadman found themselves cast adrift atop one large chunk of ice as it broke away from the cliff; one seaman lost his balance and went into the water, but all were unceremoniously hoisted safely back on board in a rope cargo net, fortunately suffering only from bruises.

Scientist takes the plunge

In my early days as a junior officer, I fell in the water a few times myself, nearly always from the boat when attempting to land on icy rocks. However, the most miraculous survival of someone doing so, was that of a 60-plus professor of cartography, on a summer visit to oversee the groundwork for the aerial survey of the Dependencies. Having been ashore, he was returning in the motorboat, and against all the rules he stood on the foredeck holding on only to the painter,[54] leaning back as he did so, as the boat weaved between comparatively small ice floes, when it hit some ice and stopped abruptly. He shot over the bow and disappeared under a floe. Despite immediate efforts by those in the boat, it took a full 20 minutes before he could be retrieved and brought aboard, and he was thought to be dead. He fortunately was not. There must have been an air pocket beneath the ice which he had used to avoid drowning, but he had also defied all the accepted concepts of time it was thought possible to survive in the near-freezing water, usually quoted as about five minutes. The boat then had a run of about four minutes before he could be brought aboard, given a small brandy and a cup of tea, and spend some time in a lukewarm bath before being wrapped up in his bunk with hot water bottles. Even at this latter stage we feared for his total recovery, but at a few minutes to midnight, for it was Hogmanay, he appeared at the wardroom door in his dressing gown to say 'Happy New Year'.

Life in command

My job was totally absorbing. I rarely had any time when I was not either actively involved

---

54 Line attached to the stem of a craft.

or keeping a watching brief over events. In difficult ice I would spend every hour possible conning the ship myself. In really bad weather, even if not actually handling the ship, I would remain on the bridge. I rarely ate properly, never having any desire to, but would be plied with toast and coffee be it day or night. On the fore part beneath the centre wheelhouse window was a ledge covered in non-slip material, which housed my coffee cup, binoculars, and a packet of small cheroots, which I puffed on in proportion to the intensity of the situation. On some sea passages when there was little ice about, there were moments for relaxation, but more often than not on such occasions there was bookwork to be tackled. If the movement of the vessel were too bad for that, I would listen to music, sitting in my armchair, and if that were not easy, I would pace within my cabin, steadying myself on various pieces of furniture, pausing to look forward from one of my windows. Music gave me much enjoyment, particularly when little else could be done I often played it to match my mood or that of the weather. It was the era of audiotapes, and I had a player lashed down upon the top of a bureau. Often people came to my curtained doorway, unable to make me hear their knocking, as I 'conducted', looking forward at the forecastle slamming into a head sea.

There were often circumstances that required my frequent, though not constant, attention throughout the night. Again, music, in the armchair, was perfect for remaining reasonably relaxed but alert enough to become immediately involved when called, or when I felt that a look at how things were going was warranted. At night if turning in was acceptable, yet sleep difficult, but resting beneficial, I had an alternative to counting sheep. I would think through some of the likely catastrophic or extreme situations that we might encounter. I did it often, with no element of worry, just running through the procedures that would be required, and mulling over all the alternatives to deal with events as they might change. There were some horrors to be thought through, such as breaking down on a lee shore or breaking down in very heavy weather in the open sea; a collision at speed with an iceberg; being beset and crushed, or fire aboard, and challenges such as towing others incapacitated or hauling a stranded vessel off a lee shore. The latter mental exercise served me well on my penultimate voyage. It became quite an obsession for me to review such possibilities when not sleeping, and now in retirement, I find I have exactly the same thoughts, but in my dreams. Mostly though, I slept well, always confident that I would be called if help or advice was needed, but very aware that it was beneficial to pre-empt any call. When I had turned in, but judged it likely that I could be needed, I would have my 'emergency gear' available. This consisted of trousers, shirt, fur boots and a padded bivouac jacket. Looking like the Michelin man, but warm, I could be up on the bridge in 30 seconds of being called.

Life was busy on a day-to-day basis, yet in hindsight each season was dominated by one or more major events.

Simultaneous dramas

Alongside the Public Jetty on the evening of Monday, 4 December 1967, I was invited to Government House for dinner. During the meal, from both the ship and the local wireless station we were informed of the seriousness of an accident that had occurred at Halley Bay, and that the Americans had been co-opted to evacuate those injured. The base was on the air providing information. I immediately returned to the ship with the governor and went to the radio room where Hugh O'Gorman, our valued radio officer, was monitoring events. We had already gleaned over the weekend that Dr John Brotherhood, the base medical doctor, and Jim Shirtcliffe, a veteran polar builder and base member of some years' experience, had fallen over an ice cliff in a white-out not far from the base, whilst out on a physiological study. They had been found by a search party that evening. Shirtcliffe had been able to erect a tent over his companion, which he had retrieved from a sledge that had fallen with them. When it became known that Brotherhood had an injured spine and a broken jaw, Fuchs in London, had requested an aerial evacuation of Brotherhood by the Americans. Swift action by them resulted in two ski-equipped C-130 Hercules being in the air and on their way from McMurdo Sound in the Ross Sea to Halley Bay the same evening. Our radio station in Stanley had already made contact with the Americans, and the airwaves were hot with instructions and advice for the reception of the two aircraft at Halley.

Although I was not directly involved, I am nevertheless including this story because it has significant drama, a humorous aspect and a nautical component. Initially it was thought unlikely that the Americans could achieve a flight right across the continent via the Pole and return to New Zealand with the casualties. I therefore set in train preparations for our departure, although it was far earlier in the season's break-up of pack ice for us normally to have sailed into the Weddell Sea. The incoming aircraft required a strip to be marked out for one of them to land. This was done by the Fids with parallel lines of empty 40-gallon drums flanking their chosen runway, and the accumulated base stock of cocoa to mark its beginning and end, with arrows indicating the direction of landing. As one aircraft circled the other landed, and when the pilot and doctor emerged from it, they were met by chocolate-coated Fids contrasting with the glaringly white landscape.

During Stanley's discussions with the Americans regarding their landing at Halley Bay, the airwaves were invaded by Deception Island base coming on the air with what appeared to be a continual recorded message: 'SOS, SOS, Deception erupting'. Believing it to be a joke, the Stanley operators gave them short shrift. Getting no reply from the base, and their SOS broadcast continuing to interfere with their reception of the Americans, they too then resorted to a taped message to the *Shackleton*. She was in the Scotia Sea, and the message requested her to proceed with all haste to Deception whilst Stanley continued its dialogue with the Americans.

Deception Island, an extinct volcano, had been murmuring for a year or more. The base Fids, through familiarity with the island's tremors, saw no need to report the disturbances. Neither did those on the nearby Chilean base, who were obtaining indicative seismograph recordings, not even when tremors began to increase both in severity and frequency and some wildlife took leave from the shores of the bay, where extremely low tides began to occur without correspondingly high ones. A month later, as we lay in Port Stanley, the island erupted in two main areas on the inner flanks of its highest peak, Mount Pond. The nearby Chilean base was overwhelmed by ash and when a large fissure was seen to be opening up towards their hut the personnel left – and, walking with great difficulty through snow laden with ash melting under their feet, they reached our base. The Chilean Antarctic relief vessel *Piloto Pardo* was the first on the scene. Having at first intended to pass through Neptune's Bellows, its master understandably balked at that, fearing a collapse of the cliffs trapping them inside the crater harbour. They remained just outside, lifting all personnel, both British and Chilean, by helicopter. *Shackleton*, under the command of David Turnbull, arrived at Deception Island a few hours later, as did the Argentinean ice-strengthened transport ship *Bahía Aguirre*, which lifted its countrymen from their base, also under threat. Turnbull, having a few hours later embarked our men from the *Piloto Pardo* in Discovery Bay on nearby Greenwich Island, then returned to Deception. Finding the wind was keeping the Bellows free from falling debris, remarkably, he entered the harbour and landed a shore party at Whaler's Bay to retrieve some belongings and pieces of equipment from the base. He then penetrated the main bay whilst the eruption was still taking place, to observe the state of the Chilean base and the effect the eruption had had upon the landscape, before retreating.

The end of these sagas was that John Brotherhood arrived safely in Christchurch, New Zealand, and quickly recovered, whereas it took two more years to resolve the problem at Deception, and then not satisfactorily. The opinion of those who knew about such matters was that the eruption had only been a minor and random event in the life of a quiescent volcano, and that a further eruption was unlikely to recur for very many years. Our base was reopened the following season. But only months into its re-occupation the island erupted again, this time at the height of a storm. Thunder rolled around and lightning struck the mountaintops, presumably partly induced by the interaction between the weather and heated ash reaching between 20,000 and 30,000 feet. The *Piloto Pardo* was again first on the scene and, with commendable flying and seamanship, lifted our men off by helicopter in a gale and a lumpy sea amidst falling debris.

After this the base was closed for good, and we were disinclined to use the inner bays of the island as anchorages; more importantly the Survey lost its airfield and hangar. I took the *Biscoe* there with some vulcanologists a little after the second eruption, who found that a 600-foot-deep three-mile-long fissure had opened up on the inner slopes of Mount Pond. More minor, but telling for our organisation, an eight-foot-wide fissure had been created right across our runway, which precluded any future use, even as an emergency landing. Our further find was that of a significant new island created within the main bay. A remarkable sight approaching the island from the east at that time was that of a multitude of tabular icebergs in the Bransfield Strait, all topped with a thick layer of dark ash carried on the westerly wind from the eruption.

Frostbitten

On a voyage to Stonington Island base in Marguerite Bay, we took the established route into the beautiful Neny Fjord, surrounded by high and steep snow-clad mountains. Then we swung to port, skirting the dangers off Neny Island, a precipitously shaped pyramid of 2200 feet, before making a tight turn to enter Back Bay, a shallow enclosure neatly situated between Stonington Island and the ice cliffs of the mainland, where the only deep water lay. By hugging them we were able to end up very near the rear door of the base; this was always a demanding and exciting but rewarding manoeuvre. Entry into Back Bay was only practical when fast ice remained within it, undisturbed from the winter, for there was no swinging room even to lie tight between two anchors. When cut into, the fast ice provided a secure berth and a platform alongside for unloading and transporting cargo.

We lay there, working easily and enjoying the situation until the onset of a gale which disturbed the fast ice. With no room for manoeuvre, I moved out of the bay and anchored close off the base in the open strait. As daylight faded and the wind increased to storm force, I laid out a second anchor. There was every sign of a Neny Howler, as the base personnel had named the tremendously strong winds that came down that fjord, another example of Föhn winds. Orographic cloud, lit with staggering hues of light grey and pink, and shot with purple, magenta and orange, poured over the mountaintops as the sun set. Occasionally I also ran the main engines to ease ahead into the wind to take the strain off the cables, but with the usual problem arising that we overran them and then, stopping, we fell off the wind and dragged.

After several instances of this, although we had plenty of room astern under our lee, the water deepened and the holding would have got worse, so I decided to leave and ride out the weather in the open sea. I had a transit on which to make the final approach to this anchorage; the base radio masts against a natural feature beyond. I used this to depart, for the wind was now beyond hurricane force, and the *Biscoe*, being a light ship, was heeling well over in the tremendous gusts coming out of Neny Fjord, which were hitting her broadside on. To see my leading marks, then astern, I had to stand out on the windward wing of the bridge as we pointed up into the wind, sometimes as much as 30 degrees, to hold our course at full speed. The degree of heel caused me to grip the steel dodger[55] on the forepart of the bridge wing to keep my footing. My gloves had become damp during the evening and although I had not noticed that it was particularly cold, when I took my hands off the dodger, the gloves and the tips of my fingers remained frozen to it, before I tore away. This was painful, for my hands were not numb; I would call the experience being frostbitten rather than a case of frostbite. As usual in the case of Föhn-like winds, when we were five miles distant from the fjord, the wind dropped away to being very moderate.

Plans drawn up for a new ship

As we wound up our activities south for the 1966/67 season, great changes were taking place in London. After years of concern that the Survey was not on a sure footing, its standing was endorsed on 1 April 1967 when it became a public body. No doubt influenced by the success of the International Geophysical Year, its international standing and the support of the Royal Society, the government Council for Scientific Policy (CSP) advised the Cabinet that it should continue to exist, and they agreed. It was also proposed that it should come under the newly formed Natural Environment Research Council (NERC). At the same time the CSP also supported the funding of a new ship to replace the *Shackleton* and remove the

---

55 Curved plating which deflects the wind up and hence over those on the bridge wing.

need for a chartered ship, which was also agreed at Cabinet level. By the time we reached Port Stanley, messages were excitedly being exchanged regarding the proposed new vessel. Fuchs, knowing that I had compiled a book of design requirements and faults to be avoided in any new tonnage, asked me to have something to present when I returned for the summer of 1967. From my earliest days in *Shackleton,* I had listed major design criteria, such as bow and stern form, double skin hull strength and integrity, through the whole gamut to apparently more minor, but in my view vital, features that should be incorporated in any new build.

As we left Stanley going north, I set about organising my notes and thoughts and putting them down on paper. I proposed that the best delivery of power from the main engines would be from a diesel-electric configuration together with a variable pitched propeller, but also that the possibility of a gas turbine booster for forcing through ice should be explored. I also thought essential the ability for a single person to control the vessel from a central console, the bridge wings and from a fully equipped crow's nest. I was presented with a formidable task, for I had notes taken over many years providing me with several hundred points. The job was made even more difficult by my wishing to draw what I foresaw as the overall design and some of its component parts, yet being hopeless at drawing. To achieve all this I set myself up on a stool at the chart room table. Not before long, a homeward-bound Fid who had been a meteorologist on one of the bases came to the bridge to take his set of the regular readings we used to compile the weather reports for transmission to the Met Office. He asked what I was doing, and in response to my answer, he said, 'I can help you.' The majority of base meteorologists, other than the most senior, were drawn from all walks of life; they were chosen for their aptitude to polar life, and on recruitment they were sent for a six-week course at the Met Office. The young man offering his services had been a draughtsman at a Scottish shipyard. For our surveying role we had all manner of drawing instruments and paper, so together we began a far more extensive and detailed project than I could ever have envisaged alone. We worked almost every day of the four-week passage home.

The return of the penguins

In Port Stanley we had taken on an unusual cargo; we had embarked about 100 penguins originating from two rookeries in the West Falklands, destined for Montevideo and thence onwards to northern hemisphere zoos. They travelled well in open-topped, lattice-sided crates, being fed on a specially prepared food by their own carer, and spending much time released onto a section of the open deck netted and secured by the crew. The Falkland Island penguins, often dirty from their life near the beaches and burrowing into the ground amongst the tussocks, unlike their pristine cousins of further south, were used to warmer weather. Nevertheless, we placed them where they got little sun and sprayed them with seawater a few times a day. They all completed the passage safely, only for us to find that there was a strike by cargo handlers at the airport from which these vulnerable birds were due to fly in cool chambers to North America and Europe. Efforts were made during our two-day stay to place them in South American zoos, for there was no question of us attempting to keep them aboard for three further weeks as we steamed through the tropics without a supply of their specially prepared food. But no home being found for them, our only option was to release them into the sea. This we did, but not from within the harbour, for a great many Montevideans would fish from the long breakwater, and we did not wish our charges to end up as someone's supper. So we anchored well off – but not far enough, because once we had let them jump overboard, the irascible little devils made it back to the breakwater. Weighing anchor, we returned close enough to watch with trepidation, as they all hopped onto it, but after a very short time, as if they were taking their bearings, and to our great relief, they dived in and swam south unharmed.

Some months later, 97 per cent had returned not just to the West Falklands, but to their correct rookeries. Before being embarked they had been tagged, so were easily identifiable on return. They had travelled 1,008 miles to the nearest point of land on the Falklands, and had then made their way to the correct island, then to the correct rookeries. The more northern waters of their route would have been completely foreign to them, and for most of their journey they would have been passing intermittently between the opposing warm and cold currents, which must have been very confusing for them. I remain astonished as to how these splendid little creatures managed such a feat of navigation and survival.

Surprises in Madeira

During the homeward passage, one of our two main generators gave us problems. Having shut it down and limped north from the Equator on the remaining engine, that too began to show signs of deterioration. Rather than risk a prolonged failure at sea, and as we were by then making good only 5 knots into the strong north-east trades, which would merge into a north-easterly down the Portuguese coast lying ahead, I decided to put into Madeira to effect some repairs. With the ship berthed alongside at Funchal, I had a visit from Risdon Beazley, a well-known marine salvor. He knew our little red ship well by sight, lying as she did each summer at Northam, across the River Itchen from his own shipyard. I gave him a tour of the ship and explained our work; he then invited me to dinner; as he was staying with friends at Reid's Hotel, it was a sumptuous affair. The following day, just prior to our sailing, he reappeared on the quay with two taxis, requesting the assistance of a seaman to go shopping! Madeira is known for its flowers, and they returned from the town laden with enough to allow all the crew to take some home. We stowed them in the cool rooms, but were nearly thwarted from getting them ashore, for the port of Southampton was full, due to a dock strike. Through our good relationship with the authorities, and my friendship with her master, both of us being Younger Brethren of Trinity House, we were given a berth ahead of *Queen Elizabeth 2*; Captain Bill Warwick slacked down her headlines, allowing us to squeeze between her bow and the dock wall ahead. For a couple of hours we enjoyed our homecoming and handed over our flowers, before moving to Thornycroft's shipyard.

Presenting plans

Shortly after arriving in Southampton I went up to the London office for a meeting on the new construction, at which I was expected to table my notes. To the surprise of everyone there, I unrolled a copious set of plans. Most of my ideas were incorporated, but one matter over which I consistently argued but still lost was with regard to the power of her main engines. I had always maintained that an ice-strengthened vessel should have a power to weight ratio far in excess of what was considered the norm. I held that the extra power could be utilised to force in unbroken or heavily concentrated broken ice without fear of damage, to assist in the very conditions that most held us up. My director and many other laymen in the matter of working pack ice, feared that more power meant striking the ice at greater speed and thus harder, and would therefore cause damage. I maintained that if a ship were put against ice at the correct speed, and the power then applied, it would not. In nearly all other respects, however, I got my way and my design drawn up in the chart room of the *Biscoe* was almost exactly as that put to the yard. The single major difference was the addition of a further deck to the midship block, to house more Fids than I had provided for. The stability took the weight of this extra deck easily, as she was exceedingly bottom-heavy in order to provide the necessary strength and integrity of the hull. The extra deck lifted the wheelhouse, and crow's nest growing above it was a bonus, for it improved the vision from both for working ice.

We had a momentary flirtation with a gas turbine propulsion booster to supplement the diesel-electric and variable pitched propeller arrangement. Tony Trotter, a serving naval commander, who had been involved with the navy's first gas turbine-driven vessel, HMS *Amazon*, left the navy to join FIDS, and to sail with me as chief engineer in the *John Biscoe* for my last voyage in her, to gain experience of our conditions. During the summer we visited Rolls Royce at Anstey to see and discuss their Proteus engine. It was an exciting concept, and met my requirement of additional power being supplied at the touch of a button when forcing in ice. However, the engine was not fully marinised, and there were difficulties in gearing the power from that unit to that of the electric motor.

Final voyage in the *John Biscoe*

I experienced a hectic summer both attending to the refit of the *John Biscoe* and being involved with the new ship, to be known as the *Bransfield*. The workload was such that I got little leave, so in order to do so, I remained in the UK for an extra three weeks when *Biscoe* sailed for her next season. It was to be the shortest voyage of my career, for not only did I fly out to Montevideo to catch her up, but I also returned to the UK from Punta Arenas in February the following year, at the end of the season south. Nevertheless I spent the whole of that season in the Antarctic, and as usual it gave rise to some notable episodes.

First, though, I arrived in Montevideo to find *Biscoe* anchored in the river more than a mile beyond the breakwater. The agents told me that because she was carrying aviation spirit she could not enter. I got them to advise the authorities that I would be taking command as soon as I got on board, and that normal procedures would then be put in place. In she came, and berthed right alongside the Custom House, as near to the city gate as it was possible to berth. I had not briefed Malcolm Phelps, who had brought her out, on the finer points of etiquette at Montevideo. Assuring the *bandidos* of my immediate resumption of duty, the release of two cases of our finest Scotch meant that the status quo was re-established – an extra case on that occasion because she was also carrying explosives! Montevideo never changed and it was the same for all of us, aircraft included, in the matter of 'entering' bribes.

As I had flown out to join the ship it seemed a remarkably short time before I was once more in the thick of Antarctic operations. It was not an easy year for ice and almost immediately a significant drama unfolded.

Beyond the bounds of probability

For several weeks during my penultimate season in the *Biscoe*, I had the pleasure of having Captain Peter Buchanan RN aboard. He took passage with us so that I could familiarise him with our region of the Antarctic, and he could observe my approach to navigation and the conduct of a vessel there, in anticipation of the navy acquiring the *Anita Dan* in 1969, and modifying her to become the first HMS *Endurance*. With Barry Dixon, the hydrographic surveyor who had conducted the Adelaide survey with us in 1962, as his first lieutenant, he brought her south during my final voyage in the *Biscoe*. Fitted with a hangar and helicopter deck, she was soon supporting the work of the Survey, undertaking her own hydrographic survey, and she pitched right into assisting with what potentially could have been a very worrying episode. But it was Peter's ambitious approach, his willingness to push the boundaries, and a disposition that both endeared him to all in the Survey and made him such a pleasure with whom to liaise and work, and consequently be very successful down south, that avoided a calamity on this occasion.

Early in the 1968/69 season the Survey took delivery of its first twin-engined aircraft, a DH 200 Twin Otter. Previously, smaller single-engine aircraft had been shipped into Deception and assembled there. On this occasion a rather more senior RAF pilot than was usually seconded to the Survey flew this larger plane from Toronto down through the Americas, across the Drake Passage, down the peninsula, and onto the strip on the ice piedmont above the Adelaide Island base. Shortly after arriving he departed for his first task south, and to become acquainted with the environment and the weather. The aircraft was carrying eight men and stores to Fossil Bluff, an isolated station beside King George VI Sound, on Alexander Island; at 71° 20 S, it was 250 miles distant from the Adelaide base, almost due south, across Marguerite Bay.

There is no doubt that some of those aboard were there for a 'jolly' and were not essential for the task in hand. When they were almost at their destination, the Bluff reported worsening conditions there, and it became evident that a deep, tight depression was approaching at speed from the west. The pilot decided to abort the mission and return to Adelaide Island. As the wind rose and visibility deteriorated on account of snow, they climbed above a low cloud base, but then lost the signal from the Adelaide homing beacon. Conditions worsened and they became more uncertain as to their position, and as fuel was running low the pilot decided to put down on the sea ice of Marguerite Bay and await an improvement in the weather. He mistakenly thought that they had glimpsed the high mountain beyond the headland at the entrance to King George VI Sound, so erroneously fixed both his position and the required course for Adelaide Island.

We in *Biscoe* became aware of the incident through Don Parnell, the base leader/wireless operator at Adelaide, but became concerned when he advised us that he had lost contact with the plane immediately on landing. Had it crashed, perhaps coming down on uneven sea ice? The last report from it had been, as it descended through thick cloud to below 300 feet, that it had glimpsed the ice and was landing. Anticipating an involvement, we had already moved south from west of the Biscoe Islands; and with Adelaide Island in sight, Hugh, our radio operator, tried a range of frequencies in an attempt to establish contact with the plane. I needed some bearings, utilising our radio direction finder to establish the plane's position, and took the first as soon as possible, believing then that the plane was almost certainly disabled. I needed to know its position to make a decision as to where to enter the Marguerite Bay pack ice to give myself the shortest passage through it, to reach either the downed aircraft, or the edge of the fast ice it had possibly landed on. Overhead assistance in the search for the aircraft from the single-engined Otter at Adelaide would be available when the weather cleared, but I had become accustomed to the navy not flying beyond 25 miles from their ship in these waters, so I discounted help from HMS *Endurance*.

She had however been alerted to the problem and was heading south. Don Parnell had been able to take a single bearing by radio during the plane's last transmission, but despite that being a significant achievement, the bearing looked wildly implausible. Hugh and Don persisted in trying to make contact on various frequencies and ultimately succeeded in hearing a carrier wave[56] but no voice from the aircraft in their receivers. Assuming those on the ground could hear them, they advised them to tap their microphone repeatedly, which they did, proving that they could hear them, and enabling us to take a further radio bearing as we progressed down the coast. This crossed with the first, positioning the aircraft beyond the mountains of the peninsula and on the shelf ice bordering the Weddell Sea. We became very concerned. Either our bearings were incorrect or the aircraft had crossed the peninsula, the latter being most unlikely, for that spine of mountains rose to over 6,000 feet. Hugh and Don then devised a series of questions that those aboard the aircraft could reply to by tapping once for yes, twice for no, and three times for don't know. At first those in the plane were disinclined to play our operator's game, not being aware of their own plight, but which was then dawning upon us. As 24 hours passed we took more bearings, which crossed perfectly and confirmed their position on the far side of the peninsula mountains. At the same time, with the weather clearing to the east of the mountains, the sun had come out. Asked about the ice they had landed upon, where the sun was and its relationship to the mountains that they could now see, it became certain they had indeed landed beyond the peninsula, somewhere on the Larsen Ice Shelf, having been blown over and beyond the mountains, probably through a pass of at least 4,000 feet, and very lucky not to have met them or let down earlier and crashed amongst them.

The plane had not been damaged on landing, other than to partly knock out the radio, and now, told of their exact position, they took off and flew 20 miles to search for an old food depot that Fuchs in London had, via ourselves, told them of. They could not find it, and the plane,

---

56 A fundamental electromagnetic wave producing a basic sound until modulated for frequency and amplitude to carry a signal (speech).

lacking enough fuel to fly back to Adelaide Island, landed again, not far from their original position. We now had a different problem, and one we could do little about. *Endurance*, by that time outside the ice of Marguerite Bay, confirmed that they had found the aircraft's position, as we had done. Peter Buchanan was asked to devise helicopter assistance if possible. To overcome the restriction placed on the distance that the helicopters could fly beyond the ship, he positioned the *Endurance* so that his two helicopters could fly into Adelaide Island with fuel for themselves, and then onto Stonington Island with yet more, in order to refuel themselves again and mount a rescue. Four times from there they attempted to reach the Otter, inhibited by their payload in getting over the mountains, and icing up in the low cloud. More than once, those on the ground could hear the helicopters returning overhead when they were unable to let down. Finally the helicopters succeeded in delivering fuel to the stricken plane, and all three aircraft returned to Adelaide Island base. Fourteen days had elapsed since the crash landing, initially ones of concern, then of bewilderment, then of further concern, before a very satisfactory result, the latter entirely due to some very determined and professional flying by the navy pilots, and Captain Peter Buchanan's[57] vigorous pursuit to bring the incident to the best possible conclusion.

Huskies

There were no dogs aboard that downed aircraft on its trip to Fossil Bluff, as there might have been had the flight been taking a field party to some exploratory campsite. Then there would have been some further complications. We carried dogs from time to time, often transferring them from one base to another if a base closed or a new one opened. Our Antarctic huskies had their origins in Labrador and Greenland, 20 having been landed from the relief ship at Hope Bay in 1948. There was a strain of wolf in their breeding, which was important in several respects – particularly their strength and toughness, allowing them to survive outdoors in the most ferocious weather, and their pack instincts. The latter was essential to their running as a team when sledge hauling. In a team usually numbering nine, a king dog would evolve as the dominant animal, stronger and heavier than the rest, somewhat of a bully boy and taking no truck from the others, but rarely the most intelligent of the group. That might well be the lead dog chosen by the team's handler, often a bitch. Whilst the often lighter lead dog ran at the head of the span, the powerful king dog would most likely be at its rear immediately in front of the sledge, pulling most strongly when hauling. When not hauling on field trips they were spanned out beyond the camp. Similarly at a base they would be shackled to a wire, where they would be fed seal meat, and just unable to reach each other.

Carrying them aboard, we replicated that, and *Biscoe* could comfortably accommodate a team on each side of the boat deck. The dogs would lie contentedly in the sun or curled up against the snow, but sprang to life and howled when we shuddered on hitting a piece of ice. Sometimes we would carry more than two teams; once there were 80 dogs spread out all around the decks. On an evening after their feed they would often put on a chorus of howling, the teams taking it in turns to respond to each other, mimicking their behaviour ashore. Getting

---

57   Later Vice Admiral Sir Peter Buchanan.

them on board was an amusing problem. Bundling them up Jacob's ladders[58] was awkward and took some effort; putting them in rope nets to be lifted on board by the derricks was not satisfactory, for they tumbled together and fought; but placing them, tightly bowsed down on wooden trays by their collars so they could not fight, then lifting them aboard, worked well but was slow. The best method we evolved to bring them aboard was to put them in a lifeboat, even getting them to jump into one lowered down onto sea ice, in which they could be lifted up to the boat deck.

Most Fids had a sentimental attachment to the dogs, particularly their own teams, but there was more to it than that. The relationship between man and working animal is always special, but in the Antarctic the relationship between those sledging and their huskies is unique. There is a reliance on each other for survival; the man cares for and feeds his dogs, and digs them out when they are snow-covered in a blizzard, whereas the husky has a sixth sense as to the dangers of crevasses, thin ice and open water up ahead when crossing sea ice, whilst having a light bearing upon the ground through its paws, relative to its weight. There are innumerable tales of huskies warning their masters of perils of polar travel and even saving them from them. Also of sad but extraordinary stories of behaviour by these splendid animals, such as when three men were lost travelling over sea ice that, presumably unexpectedly, broke up on their sledge journey between Horseshoe Island and the Dion Islets in Marguerite Bay; this event was very poignantly in my thoughts some years after the incident, when we managed to reach those isles and erect a cross in their memory. About one week after the silence giving the first indication of the tragedy an exhausted dog from one of the 14 in the two teams returned to Horseshoe Island. A fortnight later, more dogs reached Stonington Island, and over a month later some returned to Horseshoe Island. In all, nine of the fourteen survived. Significantly their harnesses and traces were not only chewed but cut clean through, undoubtedly by their handlers allowing them to escape from the accident which had befallen them. They had then traversed the sea ice, presumably jumping and swimming between floes to reach the safety of land. No act displays better the bond between the Antarctic sledger and his huskies than that of cutting free his canine companions when in such mortal danger himself.

An altogether happier tale was one which was the result of that difficulty of getting dogs aboard, before we developed the boat lifting method, and was on everyone's lips once when we had arrived at Horseshoe Island. It is a tale of a single dog's lone accomplishment. At the time that the base on Detaille Islet was being closed down, the ship was loading the dogs when one escaped and ran off. Despite a chase it could not be caught and with much distress it had to be abandoned when the ship departed. Three months later it came bounding into the Horseshoe Island base, having negotiated a convoluted route over land, glaciers and sea ice of 60 miles in winter, with little chance of having found penguin or seal to eat en route. To have initially taken the decision to leave its home base, which it must have first returned to in order to live off the seal pile, and then taken the decision to set off – in the correct direction to rejoin

58 A hanging ladder consisting of two ropes supporting wooden rungs or steps, which can be stowed rolled up, ready for slinging over the ships side.

its companions both human and canine – was astounding, particularly when the dog had travelled that route only once before during the previous winter; no tracks could possibly have remained. Scent? Maybe despite the previous tracks being buried by at least two feet of snow the dog's nose could still have discerned an element of it – but there still remains its decision to leave the safety of its food pile to venture into the wilderness in order to regain that bond with its colleagues, human and canine.

Night of terror in a polar low

Towards the end of this season, with far less daylight in each 24 hours, we had left the Argentine Islands, and I struck west through the French Passage to clear the dangers of the Biscoe and Pendleton Islands before shaping a course south towards Marguerite Bay. By late afternoon the barometer was falling rapidly and the wind freshening. I had the lashings of anything carried on deck checked, and took the precaution to pass additional wires over the motorboat and scow on the foredeck. In late summer, with much of the pack dispersed in this area, the fetch over which the sea and swell built would often extend into the Bellingshausen Sea; at similar times in past seasons I had experienced huge swells in this region, together with a nasty cross sea from the shifting winds of fast-moving depressions. I also remembered the loss of our scow and the two base generators stowed within it, that we had lent to the *Kista Dan* a few seasons prior, in similar conditions to those expected; it had been washed overboard in these waters. We could not afford to lose ours; this heavy, slab-sided and ugly but most vital piece of cargo-carrying equipment, together with the workboat, were the mainstay of our discharging operations when at anchor.

When darkness fell the temperature was found to be dropping fast, and as we hove to conditions began to border on the extreme. With the barometer at about 970 millibars, the swells were large but immeasurable in the dark, and the wind began gusting to 90 knots.

I had gradually come to handle the vessel through most swells by climbing each face of water with enough way not to fall off and broach,[59] taking the full blast of wind atop each beast and then easing the power as we descended into the next trough, trying my often fruitless best not to bury the forecastle in the depths between the peaks, where an uncanny calm prevailed. Ice started to form on many surfaces such as rails and ladders. There were several major icebergs around, but none of sufficient size or shape to take shelter behind. Yet in that seaway there was much calving taking place, producing an abundance of bergy bits the size of a cottage, innumerable growlers and much brash. With each pounding of the bow, solid water was flung onto the bridge front, frequently reaching the windows, with the continual concern that a smaller piece of brash, lifted within the water, would be thrown against a window and break it. If that window were adjacent to the controls the situation would have become very difficult or worse, the inflow of water knocking them out.

Turning to run before the sea, whilst an attractive option to reduce both the formation of ice and likelihood of shipping brash, was not possible; I did not fancy running before

---

59 Be driven around to present the side of a vessel to wind and sea and then to have the lee decks pushed beneath the surface.

such a tremendous swell for fear of being pooped, regardless of which islands with off-lying dangers lay due downwind, and we had no radar. For the first time in my experience the radar scanner had stopped turning, due to the wind having risen to hurricane force. In the dark, the growlers and bergy bits could be seen only when there was almost too little time in which to take action, and difficulty in taking it when one had seen them, for as they loomed ominously out of the driving snow the bow plunged down upon them rather than reacting to any helm. Then, passing down the ship's side only fractionally off, they quickly disappeared like phantoms. I feared that, being near the centre of this depression, the wind would back as the barometer rose, causing a nasty cross sea as the storm gradually blew itself out; but the barometer continued to fall. We had no option other than to remain hove to, and hope that no ice would hit our wheelhouse windows or that we would hit a bergy bit and the build-up of ice on the topsides could be contained. Time seemed to stand still, though the hours passed as we dealt continuously with the same set of problems, nerves stretched trying to anticipate the next move to counter a tremendous swell or avoid a piece of ice, whilst mentally grasping for any sign of change, hopefully improvement. The barometer reached a low of 928 millibars according to an officer in the chart room, most of them being up and about on the bridge by then. It was great to be supported by a team that just appeared on the scene to assist without having to be asked.

Then the incredible screaming of the wind, which we had strangely become accustomed to during the long hours of its ferocity, suddenly stopped. An eerie calm enveloped us, the sky cleared and the moon irradiated the whole scene with a luminescent light. As we realised that this was the eye of a storm, the ship's motion worsened; the sea, not running steadily with the wind, momentarily lost its direction, and I lost my ability to hold the ship on a course into the weather to improve her motion. Although it was expected that the wind would shift as the barometer rose, the wind came in violently and without warning from almost the opposite direction from before, tearing into the swell that now opposed it. The impossible question arose as to how to quieten the ship down, ship less water and make her safer as we were battered from all sides by a totally confused sea.

For some time we were able to head the original sea and swell, but with the wind rising above 120 knots, the highest possible reading on our anemometer, it was soon impossible for us to hold our course. The sea, still terribly confused, continued to fall about in all directions, until the new wind got a grip, stripping the tops off every rise of the opposing swell, creating a layer of driven spume above the entire seascape. Such contra conditions cannot last long, and they did not. As the barometer rose, tracing a perfect V on the barograph, the cloud which had returned broke up, the second phase of snow ceased, and the sea began to run before the new wind, which dropped to a mere gale. I turned and headed it as dawn broke. Its pale light revealed that all was intact on deck, but had anything broken loose we could have done nothing in these conditions. The young third officer who had been on the 8 to 12 watch the previous evening as the storm rose, came to the bridge for his morning turn. Almost exactly 12 hours had passed, during which we had experienced not just an exceedingly violent but the most concentrated storm I had ever witnessed. As we set course once more for our destination, happily pitching and rolling all over the place, with the glow of satisfaction that comes from having dealt successfully with the onslaught of the elements, the officer of the watch going off duty asked me if he might head his comments in the rough deck log with 'Night of Terror'.

We had experienced what is now recognised and known as a polar low. This is a particularly deep, intense and concentrated low-pressure system, a depression similar in structure to a tropical cyclone with high cumulus nimbus clouds surrounding a clear cloud-free eye. They are short-lived, lasting only 12 to 36 hours and usually never more than 48. They form from convection over relatively warmer water, in our case that of the summer ice-free northern Bellingshausen Sea. They decay rapidly on reaching land, particularly ice-covered, denying them the moisture they feed on, which accounted for the quick collapse of our polar low as it hit the Biscoe Islands and then the mainland. I was to experience them again, but never one quite so sudden in its approach, its intensity, and its speed of decay.

A striking approach before a flight homeward

On our final call of the season to Stanley, I went alongside the Public Jetty, but the bow took the mud and prevented me from berthing correctly. Twice I went astern to run in at greater speed, only for the bow to re-enter the same rut, consequently preventing any swing. Twice I re-arranged the planking of the western corner and sent the bollards spinning into the sea, much to the amusement of our large welcoming party.

Punta Arenas in the Magellan Straits was my final port of that season, for I flew home from there to assist with the design of the *Bransfield*. Without a direct flight available to Buenos Aires I managed to secure one aboard an old Dakota, which put down at several oil fields en route. BA, agents we rarely used, ignored me, it being the start of a long national holiday, so I took a flight to Montevideo, where our regular and efficient agents had me aboard an aircraft to Europe almost immediately.

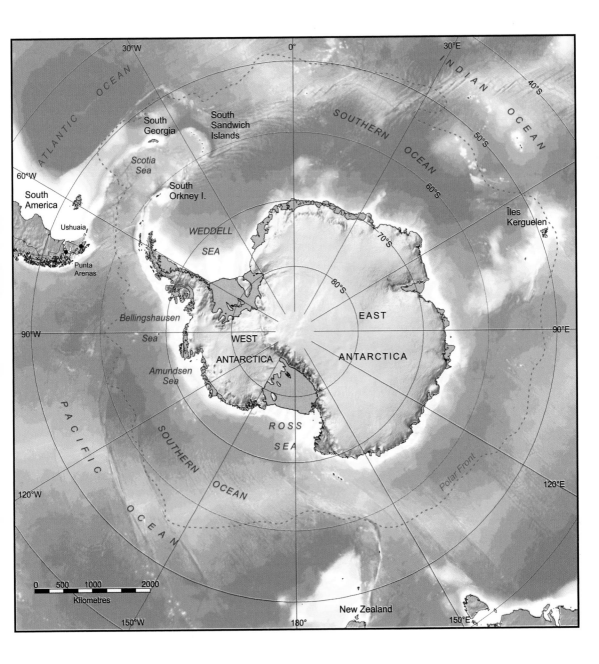

Antarctic Continent showing relationship to Australia and New Zealand, and South America and the Scotia Arc link between there and the Peninsula.

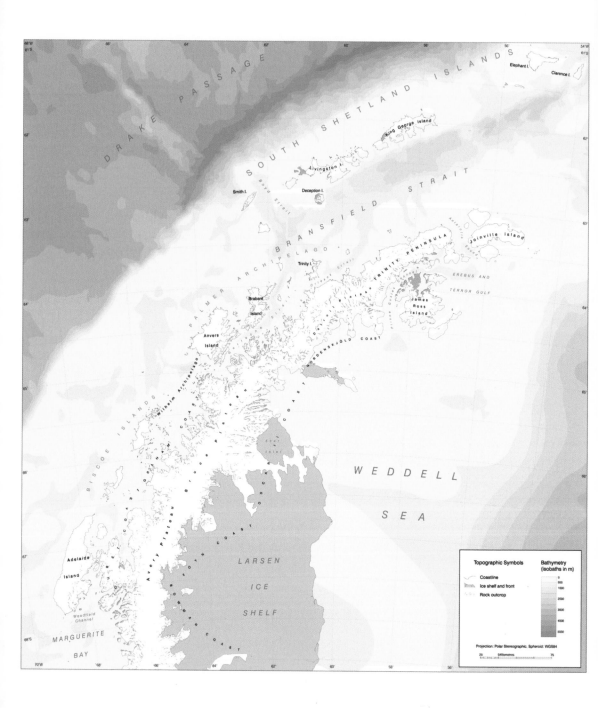

The Antarctic Peninsula. Formerly Trinity Land /
Peninsula (British) and Palmer Peninsula (American).

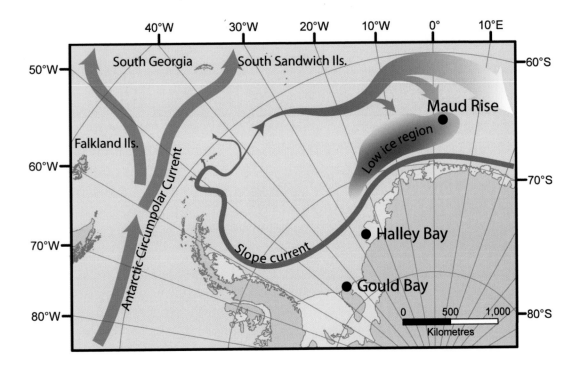

[Above] The Weddell Sea Gyre; showing the circular motion of ice and the disturbance and upwelling caused by the Maud Rise sometimes resulting in less severe ice cover.

[Below] Grytviken, South Georgia with the Shackleton memorial cross in foreground with the FIDS Base hut behind it. Government station on peninsula with John Biscoe alongside. Whaling station beyond beneath Mt. Hodges. (Crown copyright; photograph by Tom Woodfield, 1972)

Bransfield alongside at Grytviken whaling station. Hulks of whale catchers in foreground, cemetery where Sir Ernest Shackleton lies, in distance beneath waterfall (BAS/NERC copyright; photograph by Christopher Gilbert, 1979)

Bransfield, Fids, a hulk, elephant seals and King Penguins at South Georgia on a grey day. (NERC copyright; photograph by Tom Woodfield)

Shackleton surrounded by pack ice at Admirality Bay on first visit to Antarctic, 1956. Author in foreground. (Crown copyright; photograph by Tom Woodfield, 1956)

Wave peak, Coronation Island, South Orkneys. (Crown copyright; photograph by Douglas Brown, 1967)

Loose sea ice off Peninsula. (Crown copyright; photograph by Tom Woodfield, 1960)

An ice-fall on the shore line of the Lemaire channel. (NERC copyright; photograph by Dave Rampton, 1971)

Bransfield in creek at Halley Bay. (NERC copyright; photograph by Tom Woodfield, 1972)

[Left] *John Biscoe* discharging stores over fast ice with dogs, Back Bay, Stonington Island, adjacent to mainland ice cliff. (Crown copyright; photograph by Tom Woodfield, 1968)

[Right] New ice forming: the first stage in the life cycle of pack ice. Sheathbills in foreground. (Crown copyright; photograph by Tom Woodfield)

[Below] *John Biscoe* with Emperor Penguin. (Crown copyright; photograph by Tom Woodfield)

Adelie Penguins at Hope Bay. (NERC copyright; photograph by Dave Brown, 1972)

John Biscoe discharging stores over fast ice with dogs, Back Bay Stonington Island, looking towards Neny Island. (Crown copyright; photograph by Tom Woodfield)

Author eating 26th birthday cake, beset in calm off Adelaide Island, of which crevassed piedmont can be seen a few miles distant. (Crown copyright; photograph by Tom Woodfield)

Discharging oil drums from scrow utilising tracked vehicle. Author
central in pale scarf. (Crown copyright; photograph by Tom Woodfield)

John Biscoe at Argentine Islands prior to her Red repaint. With *Kista
Dan* and *de-Haviland Otter* on skis. (Crown copyright; photograph by Tom Woodfield)

[Left] *Bransfield* with *John Biscoe* alongside, showing relative sizes. (NERC copyright; photograph by Hwfa Jones, 1971)

[Right] *Bransfield* manoeuvering alongside ice shelf at Halley Bay. (NERC copyright; photograph by Charles Clayton, 1971)

[Below] Collapse of ice ramp from sea ice to shelf ice during discharge of stores from Perla Dan at Halley Bay. (NERC copyright; photograph by Eric Chinn, 1967)

[Left] Gangway landed on ice shelf, Halley Bay. Safety lines and nets yet to be rigged. (NERC copyright; photograph by Vivian Fuchs, 1973)

[Right] Author at *Bransfield* bridge wing controls working ice. (NERC copyright; photograph by Graham Soar, 1971)

[Below] Lifting men aboard *Bransfield* by crane from collapsed low ice shelf during discharge of stores at Halley Bay. (NERC copyright; photograph by Vivian Fuchs, 1973)

[Above] 10000 miles to reach this! And evidently not a metre further. The drum tower beacon that was hidden from me behind mast arriving at Halley Bay. (NERC copyright; photograph by Tom Woodfield)

[Below] Author and wife Ella with huskies on Stonington glacier. (NERC copyright; photograph by Tom Woodfield)

[Above] Author's wife Ella with party of Fids having trekked to *Shackleton* Base at 80 degrees South to clear access to it for use as a depot. (NERC copyright; photograph by Tom Woodfield)

[Below] Fids and crew off to take photographs and play football whilst beset in the Weddell Sea pack ice. (NERC copyright; photograph by Charles Clayton, 1971)

King Prnguins at South Georgia. (NERC copyright; photograph by Martin White, 1967)

Fur seal pup on South georgia. (NERC copyright; photograph by Robert Burton, 1972)

Stemming an ice-cliff to swing a mooring party ashore. (NERC copyright; photograph by Tom Woodfield)

[Above] Argentine Islands looking towards the Lemaire Channel. When mooring in Meek Channel immediately off base was not possible, an anchorage was found amidst bergs, rocks and islets, to left of picture. (NERC copyright; photograph by Christopher Gilbert)

[Below] Landing on high ice shelf in extreme cold and severe weather. Sea smoke ' boiling ' from open water beyond. (NERC copyright; photograph by Richard Laws, 1974)

[Left] Icing-up taking place in moderate gale. (NERC copyright; photograph by Tom Woodfield)

[Right] Image taken when in shelter. Further icing would have closed rails, creating a solid block of ice within endangering stability, possibly risking capsize. (NERC copyright; photograph by Tom Woodfield)

[Below] *Bransfield* with snowball on bow! Attempting to make space to berth in creek at Halley Bay. (NERC copyright; photograph by Tom Woodfield)

9 RRS *Bransfield*: Construction and Maiden Voyage

Final design and build

During the summer of 1969 I was involved to a small extent in the refit of the *John Biscoe*, but to a greater degree in the design of the *Bransfield*. In addition to a good deal of desk work, the main events of this period were a tank test for the hull and a wind tunnel test. At the former it was very interesting to observe the hull model in the tank and the trailing cotton threads around the stern indicating the flow of water into the propeller and past the rudder. A few places where the threads hung limp, indicating no flow at all, were noted, and a minor redesign of the stern configuration was proposed; however it remained very 'bow-shaped' for working astern in ice, even having a rolled iron stern post. Three parallel fins ahead of the propellor protected it from ice. However in service vibration indicated that a better water flow and protection of the stern gear from ice could be achieved, so a canopy was fitted above the propellor at a later refit. The extra thrust from this and resultant slight additional speed partially compensated for the cutaway stem of the ice-form bow, which was forecast to reduce the vessel's speed by nearly ten per cent, almost the opposite in design and effect to that of a bulbous bow. The wind tunnel proved not so interesting, all results being accepted as satisfactory. In hindsight, after a couple of seasons it became apparent that the tests may perhaps have been carried out with too much expectation of being perfect, for we suffered badly from soot and smoke from the funnel flowing down onto the helicopter deck aft. This was remedied by a tall extension to the exhaust pipes, which rather spoilt her elegant profile.

I went to the shipyard in November of that year, to be joined not long after by Tony Trotter, the chief engineer. I was supposedly appointed to the ship as soon as I had reached the UK after leaving the *Biscoe* in Punta Arenas the previous February. But neither our office nor I had foreseen the effect of the involvement of our new masters, NERC.[60] No longer were such decisions to be taken in house; the post had to be advertised nationally, a shortlist drawn up and interviews held. Amongst one other from the Survey to attend other than myself were a sprinkling of naval officers, some of whom had seen limited service in the Falkland Island guard ship, and one or two from merchant service companies such as Canadian Pacific, again with limited ice experience.

---

60 Natural Environment Research Council.

I attended before a board of eight, mainly consisting of NERC officials, one of whom was its marine superintendent; but we did not come under him, managing to keep our ship operations away from NERC's control. It was all rather a bland inquisition, none of them except Fuchs having ever been south. The marine superintendent asked what I would look out for on leaving Montevideo. He had obviously read up on a few facts, and in the Admiralty Pilot must have come across the episode of an iceberg becoming stranded on the English Bank in the River Plate in 1936! So I duly trotted out the rather ludicrous answer he wanted to hear, 'Watch out for bergs', and beginning to enjoy myself followed that with a prolonged explanation as to where and why various types of ice might be met some 1,200 miles further south. Fuchs tried to draw me out on more pertinent matters and introduced the difficulties of Halley Bay for me to comment upon.

Naively, I first said that they could all be dealt with and there should be no concern as to the continuing relief of the station. Then it dawned that he was trying to help me, and I enjoyed telling my tales of berthing alongside the ice cliffs and of its falls upon us, and of doggedly having to return to re-moor – all rather expansively, as I realised that for the majority it was the first time they would ever have heard such stories, since before BAS had become a component body, NERC had never had any interest or dealings outside the UK. They seemed as if they could have listened all day, and my interview became more like a chat around a pub fire. I got the job, but it took a further six months before I was formally appointed; by that time I had already had a small team on the ship's payroll, working with me at the yard.

Even after the adventures and excitement of voyaging south, contributing to and observing the build of *Bransfield* was an enormously absorbing and satisfying involvement. The Robb Caledon yard at Leith, part of British Shipbuilders, had won the contract. The yard was run by the younger Henry Robb, and he, his directors and the entire workforce were a delight to work with and could not have been more helpful. The workmanship of the yard was superb. The vessel was built in modules of 20 to 30 tons, firstly incorporating the double-bottom tanks which were laid on the blocks, then building up the sides, bow and stern likewise, with further modules. There were too many highlights to enumerate all, but standing amongst plywood sheets and pieces of timber as they were moved around in the joiner's shop to form a mock-up of the wheelhouse and the crow's nest exactly to my specifications was one; then seeing the form and strength of the bow, with its enormous round stem post backed by massive steelwork to which two-inch plating was scarfed, and the stern sections as they developed in reality rather than on paper were others. To see the stem running aft at an acute angle to facilitate riding up on pack ice, yet dropping vertically for its lower six feet, to give floes a thwack in an attempt to crack them and to prevent too great an amount of hull going up onto the ice and becoming stuck, was the realisation of a long-held vision. The central bridge and accommodation block was built across the Forth at Burnt Island, an associated yard, and the hull, after its launch on 4 September 1970, was towed there to receive her upper works, and then the ship returned as one to Leith. The whole build process amongst such a professional and dedicated team was an experience of a lifetime. I now fret for the loss of such expertise and knowledge with the closing of this and other similar British shipyards.

My liaison with my own office was through Derek Gipps who had come to the Survey from the Crown Agents and had the title of logistics officer. He occasionally visited the yard mainly regarding matters of extra expenditure. We saw eye to eye, and not once did we argue or fall out over any alteration to the specification or costs. My involvement was total, ranging from construction details to those relating to the interior finish, furniture, equipment and stores. Derek and I had some pleasant outings to select some of the latter three; visiting the soft furnishing manufacturers in Glasgow to select settees and chairs for my cabin and the wardroom made a nice change from crawling about within tanks to look at bilge suctions. The yard manufactured all the non-soft furniture such as dining chairs and mess tables, and that of the bridge and chart room. The joinery shop was superb, and personified the ethos of the whole yard. Among its masterpieces were the shaped desk, a davenport bureau, a dining table and serpentine-fronted fireplace with over-mantel for my cabin. Unfortunately there were setbacks, one of these being a fire in the joinery shop that destroyed a great deal of the furniture already made. Another setback, the most major causing the delivery date to slip by three months, was that the main propulsion motor fell off a low loader negotiating a roundabout on the A1, and had to be returned to GEC.

A slight dichotomy arose between my view on the delivery date of the ship and that of our director, Sir Vivian Fuchs. To have her ready to sail in October, for a full season south, would undoubtedly have been best, but my view was that she had to be finished reliably and properly with everything in 100 per cent working order before leaving. Once committed to the Weddell Sea, Halley Bay being our prime destination that first season, there would not be any assistance available for rectification of problems. Fuchs, through Derek, began to pressure the yard, whereas through Henry Robb I quietly stressed my requirements at the expense of a delay. I judged that sailing just before Christmas would allow me sufficient time to make the relief of Halley Bay before the end of the season, and that timetable was being kept to until a minor disaster struck. On an early trial in the Firth of Forth, having worked up to a good speed, a total blackout occurred and the stench of burning pervaded throughout the ship. Henry Robb appeared from the engine room, face blackened by soot. The main switchboard had fused and burnt out. We were towed back to Leith for a vast quantity of electrical equipment to be rebuilt.

Christmas came and passed unnoticed. A mad scramble to finish, which I had so desperately wished to avoid, ensued. Other areas of build had fallen behind and at one stage 28 different trades were working, one on top of the other; as bulkhead facings were put in place, someone else took them down, wishing to work behind them. New Year approached, when the port would close for three days, including the locks to the river we had to pass through before departing. On New Year's Eve we were ready to go. To set sail on a brand new ship was an exciting but somewhat complicated procedure. Various certificates from Lloyd's Register of Shipping, the Board of Trade and the shipbuilder, amongst others had to be gathered and presented to the shipping master and customs for the ship's documents to be issued and clearance, 'entering out', to be granted. At 3 p.m. I went to see the shipping master, armed with everything necessary. A delightful Scot with a twinkle in his eye, he offered me a wee dram from a bottle produced from a lower drawer in his desk. 'To christen your new ship,' he said. Then, of course, we had to toast the crew and myself, and so once or twice more we raised our glasses. Eventually he said, 'You'll no be wanting to sail on Hogmanay, Captain! Have another wee dram.' With the locks closing at 5 p.m. for three days, I could visualise the itinerary of our maiden voyage disintegrating before my eyes, so I escaped his rascally generosity, to get back to the ship clutching my briefcase containing the requisite precious documents, and we left the shipyard berth. In the locks the whistle jammed and we made our final departure from Leith with it continually sounding as the gates opened; we entered the Forth at 4.50 p.m.

We arrived in Southampton immediately before a weekend but loaded throughout whilst innumerable visitors came aboard. On Monday, 4 January, 1971 we sailed for Antarctica on what was to be a very memorable voyage.

Maiden voyage

The voyage south was to omit the peninsula, and head via Montevideo, the Falklands and South Georgia straight into the Weddell Sea. The relief of Halley Bay was still feasible, just within the accepted seasonal window. As we left the Canary Islands astern and the weather warmed, unusually high engine room temperatures were experienced. Quickly we realised that the engine room extractor fans had been fitted upside down, and this took several days of very hot and difficult work, rolling in a seaway, to rectify.

It was the engineers who suffered from the consequences of that fault, but the next to manifest itself affected all on board. A daily increase in the amount of roll commenced in the more lively weather of the northern latitudes, but little notice was taken of this at first. In the long, low swell of the tropical latitudes, we expected some rolling, because in order to reduce friction and obviate any snagging when working ice we had no bilge keels, and the welds between plates on the hull were buffed. However, each day the roll increased, eventually to dangerous proportions. It required little deduction to realise that something was very wrong, and the fault almost certainly lay with our three passive stabilisation tanks. They were fitted one above the other, amidships against the forward engine room bulkhead, across the breadth of the ship. Each was partly filled with an exact amount of water related to our loaded draught. This water moved freely across each tank as the ship rolled, except that on the centre line of each tank was a triangular weir that deflected the moving water, curling it up and then back to the side it had come from. In theory as the vessel rolled to starboard the water in each tank was directed to port and vice versa, thus dampening any roll. To enable the engineers to enter the tanks, I took a phone aft with me onto the helicopter deck, to handle the ship from there, partly because the motion on the bridge was too ridiculous, and partly because I could observe the swell better, to attempt to control the ship's motion. As the problem arose we had found that the bottom tank was becoming full, the middle tank filled above its proper level and the top one empty, and assumed correctly that the upper ones were leaking into those below. When they had been drained it was found that although the upper two tanks had their own air pipes to port and starboard, they vented into one larger common pipe from the lower tank, down which water was leaking. That was the problem understood – but how to rectify it was another matter.

The third engineer, Larry Buchanan, a genial and immensely practical lad from Lancashire, came up with the idea of inserting individual plastic vent pipes to each tank rising up to deck level inside the existing larger steel pipe from the bottom tank. That was how they remained for the rest of the voyage and the stabilisation system was then deemed a great success.

Our next problem was minor, and one that is often seen on ships. We had water coming out of the light fittings in the corridors. Water pipes were invariably run along the corridor deckheads and some joints where they had been fed off to enter cabins had worked loose. The water had got into the conduits carrying the cabling and exited at any break, such as for a light fitting. It actually did not cause any great problem other than an inordinate amount of time being spent taking down deckhead panels to find the source of the leaks.

A conundrum that was to be solved by Hugh, the radio officer, one of several men who had come with me from the *Biscoe* to the *Bransfield*, was at first amusing until we realised that someone could have an accident. The ship was fitted with heavy fire doors that were held open by magnetic catches. When they were to be closed, a switch was thrown on the bridge, demagnetising the catches and releasing the doors. Our problem was that they would all suddenly close on their own from time to time without any warning. Often making an ocean passage there are a few pleasant rituals such as tea and toast on the bridge at 0630 hours for the officer on watch, when the stewards turn to, and tea and tab-nabs (buns) at 1500 hours for the afternoon watch keeper. Unfortunately for a mess boy, some others and I had assembled on the bridge to discuss our programme during the afternoon watch. He ascended the bridge internal companionway with a trayful for us all. As he reached the top step the fire door closed in his face and he and his tray were sent tumbling down. There were many similar instances, and we were hard pressed to find the cause. Eventually by some clever deductive work Hugh discovered that every time he transmitted on one particular frequency, used regularly to send met reports, the fire doors closed. The cabling from his radio office abaft the bridge to one of the aerials had been laid close and parallel to the cabling from the fire door master switch. Induction occurred, current jumping from the aerial cabling to that of the fire doors, releasing them. More deckheads came down for remedial re-routeing of wires to be undertaken.

Bolts sheared off within the electric motor during the outward passage, but did not ruin the windings, so a catastrophe was not caused. Our maiden arrivals in both the Falklands and South Georgia were rather more dramatic than I would have wished. Having run up Stanley Harbour at several knots, just as I swung to head for the jetty where the *Biscoe* already lay, the electro/hydraulic combination control of the engines failed. Anticipation of seeing the new ship arrive was high, and there was quite a crowd lining the front road to watch. To those ashore we were apparently making an initial curtsey to our home port and sister ship, as the mate readied the anchors for dropping before we hit that infamous sewage pipe, whilst with a good deal of sweat on my brow I altered course slightly for the shore to avoid hitting the *Biscoe* all the while letting the engine room know in no uncertain manner that I needed them to give me full astern. This they did just in time, and restoring control on the bridge I brought her up short of the *Biscoe*, and with some apparent élan, slid into my berth. A very lucky outcome to an incident that could have been embarrassing at least, disastrous at worst.

About a week later, before swinging into Grytviken harbour and still at full speed, with Hobart Rock, an isolated danger right ahead, the same engine control failed. Able to avoid the rock by altering course into the harbour, we went in still at full speed, and made an exaggerated loop whilst the engineers again rectified the problem, after which I managed to reduce speed and cautiously berth. We were dressed overall,[61] as we had been for our maiden arrival in Stanley, and once again none ashore had any idea of our problems. Rather the reverse, for I was gaining an unintentional reputation for some showboating with my new ship.

The ancient wooden jetty at King Edward Point was deteriorating badly so, as with the *Biscoe*, I chose to moor my much larger ship near it rather than alongside, for fear of damaging it further. We attached a wire from aft led along the ship's side, then with plenty of slack, attached it to the anchor. Dropping this square off the jetty we paid the cable from forward and the wire from aft, putting out mooring lines from both bow and stern to the land so that we brought up parallel to, but two feet off the jetty. The crew tended the after wire carefully and the manoeuvre was a great success although during its execution I was worried that the wire would foul the propeller. Normally this rotated constantly whilst the pitch was varied, but on later similar occasions I thought it wise to stop it completely between manoeuvres.

Personnel

Prior to our passage into Halley Bay, our chief officer broke his ankle and had to be replaced. The new officer, together with some of those officers already aboard and some later young recruits, were to form the core of the Survey's marine team for the next several decades:

Chris Elliott had joined me for my last voyage aboard the *John Biscoe* as her junior deck officer. He then stood by *Bransfield* in her latter stages of building at Leith, and sailed on this maiden voyage as second officer. He was an extremely keen, thoroughly likeable young man, easy to get along with, and above all an excellent seaman. His aptitude for handling small craft was frequently watched in awe by those on board, as he lifted shore parties with inflatables in storm-force winds on dangerous coasts. During the same period he was developing his skills as a ship handler and navigator.

John Morton had joined me in Leith as chief officer and sailed in *Bransfield* as such on the maiden voyage. After he broke his ankle, Stuart Lawrence joined us to work in parallel with Chris as joint chief officers/second officers, for we knew that Chris would be taking leave immediately after the voyage to sit his master's certificate and participate in the single-handed transatlantic yacht race. Stuart's arrival was of some concern to me, and of much mirth to both of us later. He had been recruited from Canadian Pacific to fill a position as the most junior officer on the *John Biscoe*, and when John Morton had to go ashore, the director proposed sending Stuart as his replacement to the *Bransfield*. When I was advised of this, I sent a rude message telling the director to find appropriate personnel for our ships; Stuart got to know of this, but on arrival it transpired to my surprise that he was not as inexperienced or junior as

---

61 A line of flags rigged from right forward, to the top of the mainmast, thence to the stern.

I had thought; not only had he got a master's certificate but also he had previously sailed in a much more senior position than he had applied for on the *Biscoe*.

Whilst Chris was quiet and steady, Stuart was a human dynamo, so this was a good but sometime fractious pairing. Small and wiry, with a sharp brain and agile mind Stuart went around the ship like a firecracker, usually wearing his trademark orange boiler suit over his warmer gear. Efficient to the extreme, he was also at times a worry because he ate so little, and not at all at lunch times when most in those temperatures ate enormously. It apparently did him no harm, and he became an excellent chief officer, taking over from me on several occasions and ultimately as master when I retired from the Survey.

At that time Chris also took command of the *John Biscoe*. I never sailed with either of them when they were master, but both now having over 20 years of successful command always well reported upon speaks for their expertise. Regrettably for them, they never experienced the freedom to operate that Bill Johnston and I enjoyed, it being steadily eroded over the years by the bureaucracy of NERC.

Hugh O'Gorman had come with me from the *Biscoe* to the *Bransfield*, as had Eric Heathorn, the chief steward, and Tony Trotter, the chief engineer. Hugh, formerly a Fid, was invaluable for his knowledge of the workings of the Survey and was central to all communications, including keeping in touch with all field sledging parties. When not working almost incessantly at keeping everybody abreast with what was going on, he assisted me in the personnel management and most clerical tasks.

Eric Heathorn had risen from being the cook in the *Shackleton* to chief steward in the *Biscoe,* and then *Bransfield*. He was thorough and exacting, and provided for those on board, plus many others visiting from the bases, every day of the voyages of up to eight months. He fed us so well that I never heard a single complaint about food, in fact often quite the reverse. At the rebuilding of Halley Bay he was catering for 150 people in two shifts around the clock; they ate vast quantities of meat and potatoes to provide the calories required for such hard work in low temperatures. He ran my bond of liquor, tobacco and the canteen, all of which had to be thought through and ordered prior to the voyage, as did the food stores of meat and vegetables, fresh, dry and frozen, not just for the planned voyage but also for the eventuality of overwintering if we became beset.

Tony Trotter was a very capable chief engineer, and invaluable on this maiden voyage. With a pleasant, likeable personality, always willing to help and give of his best, he never seemed ruffled, and ran his team authoritatively yet with a gentle hand.

First voyage in *Bransfield* to Halley Bay

The voyage to Halley Bay was made in good time, for it was a reasonable season for ice, yet there was sufficient heavy ice for the vessel to show her abilities. Handling the ship from the crow's nest was extraordinarily beneficial, particularly when going astern, the amount of dead water[62] being so very much reduced. The vessel could be steered astern avoiding the heavier

---

62 The area of sea obscured from sight by the after structure of the vessel.

pieces of ice, like positioning the bow when forging ahead. That speeded up the backing and filling required in heavier ice. The added height gave far greater vision ahead for sighting leads and planning the immediate route, but in conditions where no sternway was required, I still preferred to be surrounded by my support and in touch with all that was going on, and con from the bridge rather than being isolated in the crow's nest.

Her ability to force and break was proven, as was the benefit of the short vertical stem to prevent any length of the hull riding up too far onto the ice and our becoming stuck. Various modes of traverse through the ice were always the norm, but the exciting difference in the *Bransfield*, with her combination of diesel-electric power and variable pitch propeller, was in our ability to apply that power and the consequent method of progress. The watt meter was the instrument that transformed the handling of the ship in ice. Charging the ice from a stationary position, having backed astern to provide a short run ahead in fairly open water, we could adjust the propeller from an almost feathered position to a much coarser one as the vessel gathered way. With the shaft revolutions being self-adjusting, it was simple to obtain the best possible output, and therefore thrust, merely by increasing the pitch of the propeller whilst keeping the needle of the watt meter as high as possible, yet out of the overload sector. Furthermore, whenever the vessel rode up on pack ice, we could by watching the watt meter increase the power to maximum, with the stern fractionally but sufficiently kicked up to provide a chopping action. Thus a progression of making way through the ice, of charging and riding up onto it, then applying maximum power to chop and weigh down on it to open a way forward, was achieved. This could be repeated again and again, all managed by my left hand as my right did the steering.

As we progressed through several hundred miles of varied pack ice, denied clear skies, and with ice cliffs, bergs or pack despoiling the horizons needed for us to measure the altitude of sun and planets, we perfected our dead reckoning on icebergs. When we arrived at Halley Bay, there was excitement on the bridge when the marker on the cliffs identifying its location was sighted. Sighted, that is, by everyone save me. The Fids had built a tall tower of empty black 40-gallon oil drums to guide us in when making our landfall from offshore, rather than when progressing down the ice cliff coastline. Only when I moved away from the control console, off the centre line, did I see it. We had raised their slender mark dead ahead behind the topmast, after working in ice for a week since our previous astro-fix!

The relief of the base was fairly normal. Because it was late in the season, the ice ramp from the sea ice up to the shelf ice had disappeared at sea level, forcing us to lie alongside its remnants right at the very head of the finger creek. The space to berth there, between the opposing cliff faces, was about 20 feet too short, and made for some amusing chipping with the bow to get in. Once alongside our red paint on the cliffs bore witness to the tight squeeze, whilst the giant snowballs on the prow did not do much good to any of the lighter fittings there.

Record southern latitude reached

After the relief we sailed south some 200 miles to review the state of Shackleton Base for use as a depot. Sir Vivian Fuchs, who was with us at the time, had made his base camp there, the starting point for his trans-Antarctic crossing in 1957. We then ventured slightly further south again, through moderate ice, into Gould Bay, reaching the farthest point south attainable by a vessel in the Weddell Sea. With our bow into the ice cliff and steaming gently ahead to hold our position, Chris Elliot and I took sights at midday to obtain our latitude. We could see the sun, but only occasionally a horizon, for dense sea smoke was coming off the water, frequently obscuring it. Measurements were only made possible by some intermittent gaps in the smoke, presumably caused by the fluctuating wind.

Sea smoke is met in the highest latitudes of polar waters; it is formed when very cold air, such as that blowing off the continent to our south, flows over a relatively warm sea. Evaporation takes place from the sea faster than the air can absorb it; the water vapour produced condenses in the form of fog rising from the sea, a phenomenon similar to steam rising from a boiling kettle. With our air temperature below minus 35 degrees C and the seawater around us at minus 3, the resultant fog was dense, and this was made worse by our propellers churning up the seawater astern, increasing the rate of evaporation, so that at times I resorted to having the propeller stopped momentarily to reduce the sea smoke and improve the clarity of the horizon.

We had taken as many altitudes as possible around noon. The midday altitudes were standard observations for us, but not with the sun so low and our fingers too cold to adjust the verniers[63] on our sextants. For seamen this was a unique experience, though not so for anyone trekking towards the Pole. We corrected our collective sights from the nautical tables, the adjustments for such low altitudes in high latitudes being particularly large, coming to conclusions only one mile apart. When averaged, this positioned us at 78° 04' south, at the time a record achievement, but the following season an Argentinean icebreaker broke our record after the ice cliff we had stemmed against had calved and broken back several miles.

The sea smoke prevailed intermittently for some days on our voyage to Gould Bay, and yet again northward, almost to Halley Bay. It was a frustrating, indeed quite alarming inconvenience, because in areas of open water such as leads and pools between the pack ice the smoke rose densely, as if we were in shallow fog. Sometimes one could see above it from the crow's nest, which was interesting in that from there icebergs could be seen, but this was not the slightest use in dealing with ice at sea level. What was extraordinary was that when we could not see this meant that there was water with very little or no ice coverage and therefore we could proceed at a good speed; but whenever visibility cleared we would need to go full astern to stop, for ice lay ahead in our path. This alternating very poor visibility and good visibility, with sunlight suddenly re-appearing and glaring off the ice, became very tiring. My eyes got so weary that at one point, when I reversed the pitch of the propeller to go astern, accompanying this with yet another expletive and exclaiming that I had been surprised by the rearing piece of ice shelf ahead, the chief officer quietly said that it was only a large flat expanse of new ice, less than a foot thick, easily cut through, and asked if I would like him to take over. I accepted.

---

63 The fine control to measure the minutes and seconds of a degree of altitude.

Beset in brash ice

With everybody tired after hours of dealing with sea smoke, I lay to for that night instead of driving further north. We had passed a great deal of fast ice attached to the ice cliffs to our eastward, with distinct bays on its seaward edge. I chose one such bay where some loose pack had gathered, and lay amongst it quietly and comfortably, but I avoided going right alongside the fast ice edge for fear of being pinned broadside against it and unable to work my way off it in an onshore wind, even though the latter would have been unusual. Unexpectedly, though, a light breeze came in from the west and more ice but mostly brash driven by that wind, filtered itself amongst our adjacent floes. Overnight the brash consolidated, and by dawn it filled the bay between the floes. But the severe night temperatures had solidified it and we were bound in, every inch of the hull frozen down to four or six feet below the water line; I could get no movement whatsoever when I attempted to go ahead or astern. On our attempts to roll the vessel, she still remained firmly in its grasp. My concern mounted, for it was towards the end of the season, almost as far south as it was possible to voyage, the temperatures had plummeted, and the wind was not the expected helpful offshore one. How embarrassing after such a successful maiden voyage, achieving the record south, to be forced to remain there over winter. Patience was eventually rewarded, however, and after many hours spent going ahead then astern, ranging the rudder from side to side, and some limited rolling, the ship eased forward. The challenge then was to keep going through the 'porridge', but by the next day we were well under way again with a lot of relieved faces about the ship, including mine!

A split decision

Moving northward the following day with the Filchner Ice Shelf still in sight astern, we met some very heavy multi-year ice, workable only by charging, splitting and forcing, frequently going astern to make space for a more aggressive charge. Eventually, though, it became too much for us, and although we were not beset I stopped, awaiting a better breeze to loosen and move the ice about a bit, for it was almost calm, and would certainly change within hours. I told Stuart to let anybody who wished to go down onto the ice, to walk, take photographs of the ship and penguins, or play football.

Almost the first to go off was the chief engineer. I watched from the bridge wing, and he seemed to hesitate going down the Jacob's ladder. When he reached the ice some ten feet down from the forward well deck he stood still and stopped anyone else from coming down the ladder. I thought, 'there can't be anything wrong with that ice; you could safely drive a tank across it', and there was not an inch of water to be seen. He returned to the deck, came to the bridge and quietly told me that we had a six-foot vertical crack in our shear strake. This was the uppermost strake of the hull plating of the vessel adjoining the main deck, and one of the heaviest, being two inches thick in the *Bransfield*. The crack, about an eighth of an inch wide, could easily run completely around the hull. The possibility of the ship being zipped into two parts was an immediate and worrying thought, especially as we had over 1,000 miles of ice to work until we reached open water, and then there would be heavy seas; in either of these, extra strains could occur to bring about disaster. As many stretched their legs in the sunshine of the calm evening, the engineers drilled a round hole at the top and bottom of the split to reduce the chances of it running further. This time-honoured solution worked, not just temporarily while I handled the ship gingerly at first, but right throughout the remainder of the voyage until we reached the UK three months later.

The fracture was almost certainly the result of the method used to build the ship. Once the base of bottom tanks and keel plates had been laid, the hull had been assembled by drawing together prefabricated units of up to about 30 tons. These consisted of a section of the outer hull, a parallel section of the inner skin and the transverse deep frames that together formed the side tanks. In some cases I had watched a gap of two inches between modules being closed as they were winched together during the building process. The locked-up stresses must have been enormous, with the unforeseen result that when the hull was pounded by ice, it split. As was so often the case, the welds were stronger than the plates they joined, and it was a plate that had fractured.

An Argentinean welcome

At the end of that maiden season of 1970/71 for the *Bransfield*, it had been decided that we should visit Ushuaia on our way home, because the Argentineans were interested in our new ship. They were contemplating a new vessel of their own, and a cost-effective method of acquiring one would be to replicate another one, avoiding much of the preliminary work of an entire new build. Ushuaia lies within the southern reaches of the Beagle Channel, and although this means backtracking towards Cape Horn, it is an interesting place to visit with an unusual approach, so nobody minded too much about the detour despite being homeward bound.

As we were sailing south-westward from the Falklands the weather deteriorated appallingly. We were forced to heave to for many hours half way across to the mainland, with a storm from right ahead. Then when it dropped to a gale, we continued and passed along the southern shore of Isla de los Estados (Staten Island). It had been my intention to pass south through the Straits of le Maire separating Staten Island from the mainland of Patagonia, but it was notorious in south-westerlies for its tide rips and short, heaped seas. These were described as early as March 1741 by Commodore George Anson on his voyage to the South Seas, and in 1871 by those in the *Allen Gardiner*, which brought the heroes of the second part of this story to this tempestuous corner of the world.

From Staten Island we ran along the south-eastern toe of Isla Grande de Tierra del Fuego, with force nine gusting ten from right ahead, and although there was no swell we had a very strong current setting against us. Entering the confines of the Beagle Channel with Picton Island to port, the last 40 miles of tight navigation in the dark, with the wind still strong and always against us, following the line of the channel, entering the constriction to pass between Gable and Navarin Islands and through the five-cable-wide Paso Mackinlay – all this was demanding, to say the least. I reminded myself of where extra vigilance was required by noting the locations of known accidents, particularly where an American icebreaker had recently clipped a rock.

The Argentinean Antarctic Operation was virtually a military one, and it had been arranged that a party of several high-ranking service officers would come down from Buenos Aires to visit us, so the 18-hour delay that we had incurred was probably proving none too convenient for them. We approached the Bay of Ushuaia around midnight, and it quickly became evident that the small city of only 4,000 people was in total darkness. Furthermore, that every single navigation light was also extinguished. Perhaps from a power failure due to the extreme weather we had experienced also hitting them; but that seemed strange, for most major navigation lights, particularly those offshore, have their own independent power supply. I was in no mood to hang about, for we had all taken a terrible pasting, with great demands made of us over the last four days in almost continual storm conditions. When I entered the bay, all I wanted was to berth safely and turn in, but no pilot appeared. The place seemed dead. Cautiously I went further into the bay, and could see the outlines of some quays on the radar, but did not know which was the most appropriate for us to lie alongside. Switching on our searchlight, I made for one of them dropping an anchor as we approached, for it was fairly shallow. It transpired that the quay I had chosen was a none-too-salubrious oiling jetty.

We went alongside nevertheless, the sailors jumping ashore to take our own lines, for there was also an obvious lack of people. None to take the lines, no customs, no agents, nobody – and this was an official visit arranged by the Foreign Office. Once we were tied up, I went below, and instead of turning in I had a drink with the three 'chiefs', Tony, Stuart and Eric, to relax and discuss the extraordinary situation. About half an hour later I was told that a 'military chap' who said that he was the port captain was on the quay beneath the gangway, asking for me. I went on deck, and he shouted up 'Why have you come into my port'? To which I replied, somewhat irked and rather tired, in my best Spanish learnt in Montevideo, 'Don't be so bloody stupid! You know why we have come.' He replied that we were not welcome and should leave immediately. Sensing some South American drama, I told him that I was seeking refuge from the weather, and that neither my crew nor I were fit to go to sea again, then retired to my cabin. I also sent a Spanish-speaking member of the crew to invite Captain Rosso, for he had told me his name, to come aboard to talk if he wished to do so, but to hurry, for I was going to bed. He came up appearing sheepish, but accepted a whisky and we talked of our departing after a good night's rest. I still had no idea of what was amiss. Then gone two in the morning, a phalanx of Argentinean officers and officials arrived at my door, and I told them they were too late, for I was off to my bunk. But then in their midst I noticed a British army officer, very drunk, who turned out to be the military attaché from Buenos Aires. He went straight into my bathroom and was sick. When he reappeared he handed me a letter from our ambassador. There had been a coup d'état in Argentina; President Levingston, who had arranged our visit, had been removed whilst we were sailing across from the Falklands, and Alejandro Lanusse had been installed instead. He supported the view that the Falkland Islands should be handed over to the Argentineans, and with our Port of Registry being Stanley, Falkland Islands, our visit had been cancelled. The attaché bearing this news from the ambassador had been 'entertained' by the local military during the delay in our arrival, and then prevented from meeting us. Disgracefully, they had also turned off any recognisable lights, supposedly to prevent our entering. I said that I understood the position, and agreed to leave, but in the morning, so the posse eventually left, leaving me to sleep.

After breakfast, when we were preparing to sail, a jolly, rotund gentleman came on board and introduced himself as the British Consul. He was extremely pleasant; annoyed at the treatment we had received, well versed in the ways of Argentinean politics, and had a plan for revenge. His was an extraordinarily interesting background, his forebears being pioneers in the region. His plan to retaliate for our treatment was for him to take me on a trip inland, to view the spectacular Fagnano Lake. The depression had gone through, and it had calmed to a lovely sunny day with snow-clad Mount Olivia shimmering as a backdrop to the city; we drove around it in his 4x4 for some breathtaking views. His wife had prepared a picnic that we enjoyed on the snowline.

Having descended back into the town about 5 p.m., we went to the main hotel for an early dinner. Within hearing distance in the restaurant were several army officers, including Captain Rosso, all quite plainly seething at our presence, but not speaking to us. They had obviously been told to get us out of the port, but their diplomacy guided them not to create a public incident with the unwelcome Brits. It was the first few days of the new presidency, and I learnt later that they had been told 'to keep a lid on things'. The consul revelled in the situation. We had a lengthy dinner and lingered over coffee as our 'friends' made passes by our table, in a lame attempt to spur us on, but still without a word. As dusk fell we drove back to the ship.

Stuart met me at the gangway and I told him of my splendid day. He pointed to the nearby hillside, where there was a fort overlooking the quay, and said, 'I am so pleased, sir, because those bastards have been training their guns on us all day, and quite honestly I felt like sailing without you!' Fortunately for me, he had held his nerve. The Argentineans had put a pilot aboard early in the day for our impending departure, a cheek considering our treatment on arrival. He had been plied with drink throughout his wait, and was now legless. We sailed, without his knowing, into the Beagle Channel that was divided in that area almost centrally into Chilean and Argentinean waters. We made straight to the Chilean half, where their patrol/pilot boat was stationed, and they were only too delighted to receive the drunken Argentinean pilot from us; his own pilot boat, which had been frantically trying to catch up with us, had stopped at the median line.

From there on home it was plain sailing, so I had time to reflect upon my luck; I could well have been jailed rather than had a good day's outing, for we had never 'entered inwards'[64] correctly.

The uttermost part of the earth

I also had time to ponder over the amazing stories told by the consul of his ancestors: Captain Allen Gardiner, a naval officer of promise, lost his wife in 1834 when he was 40, and embarked upon a life as a missionary, first around the globe, but finally founding the Patagonian Missionary Society. When in 1848 he landed on Picton Island in the entrance to the Beagle

---

64 The shipping term for clearing through Customs, Health and Immigration, and being given the right to go ashore.

Channel, from a vessel trading coal to Lima, the weather was so adverse and natives so hostile that he abandoned this first expedition and re-embarked for England. In 1850 he returned to the same island, but he and his party perished, beaten by the combination of weather and the Indians.

On learning of Gardiner's fate, the Secretary to the Society, the Reverend George Despard, mounted an expedition in 1854 in a vessel named after him. The *Allen Gardiner* first went to Keppell Island in the West Falklands, thence to Tierra del Fuego. This adventure failed in its missionary endeavours in Patagonia, so Despard sailed to the Falklands in 1856 with his wife and children, two of whom were fostered. From there he dispatched the *Allen Gardiner* to progress their objectives in South America, but she was ransacked at anchor and all her crew but one murdered as they held their first church service in the little hut that they had built; Despard, dispirited, brought the *Allen Gardiner* back to the Falklands. In 1861 he sailed to England, leaving behind his 18-year-old foster son, Thomas Bridges, in charge of Keppell Island, to tackle the tasks in Tierra del Fuego that had eluded all since 1848. Thomas made frequent trips to Patagonia and brought many Indians to Keppell, learning their language and ways. He visited England to be ordained, aged 25, and on return to Keppell, sent Whait Stirling to erect a hut, prefabricated in Stanley, to establish a missionary station in the best location for their work. Ushuaia, then a native encampment, was chosen as providing the best anchorage, access to all the local tribes and space for a number of families to farm. A short while later, Stirling was made Bishop of the Falkland Islands and South America, the largest diocese in the world.

Thomas Bridges then toured England, lecturing, and met his wife-to-be, Mary Varder, of Harberton in Devon; they were married there in 1869. She returned with him, both aged about 27, to Keppell Island, where a daughter was born to them. They then sailed to Picton Island, a voyage of great hardship with a nine-month-old baby, taking 42 terrifying days – this was the same sea that we had just battled across in bad weather, complaining that it took four days! From there they went on to Ushuaia to form the first colony, and on the shores of the Beagle Channel they created a farm that they named Harberton. At Ushuaia the Bridges had a son, one of the first white natives to be born there, who wrote *The Uttermost Part of the Earth*, telling the truly pioneering story of the creation of Ushuaia, their farm at Harberton, their work with the Indians, and the notable interface with Captain Fitzroy of the *Beagle*, who brought three Patagonian natives to London. From those auspicious antecedents had come my conspirator and, almost exactly 100 years after their landing on Picton Island in 1871, our fortuitous meeting.

10 RRS *Bransfield*: Second Voyage

St. Paul's Rocks, an experience of some gravity

A small task we undertook for a season or two was to land a gravimeter from which readings of the earth's magnetic field were taken, requiring the ship to visit some remote and unusual places. For those supporting a landing it also afforded a period of spare time to have a look around, as the instrument needed to settle down before readings could be taken. On outward and homeward passages we would make similar landings, to calibrate it at sites of known previous readings. Outward bound on this voyage, one such visit was made to St Paul's Rocks. The group of 15 rocks and islets are no more than a few football pitches in total extent, the highest only 60 feet above sea level, yet their base rising from the nearby seabed 12,000 feet below. It is a megamullion astride the Mid-Atlantic Ridge, and one of only two places in the world where the abyssal mantle is exposed above sea level, forming the outcrop. In a position just one degree[65] north of the Equator, the rocks lie 600 miles from Brazil and 1100 from Africa, a tiny speck in a vast ocean. They were discovered by an unfortunate Portuguese navigator who in 1511, sailing in the *St Peter* within a fleet of caravels, hit them one night, the others sailing by. They were also visited by Darwin in the *Beagle* in 1832, who noted their utter barrenness. Then Dr Bruce of the Scottish Antarctic Expedition called in 1902 in the *Scotia*, and recounted a story of his visit sufficient to make us wary; he had abandoned any thought of landing after one of his scientists had slipped into the sea and was attacked by sharks.

Finding an anchorage was a challenge. Although preferring to lay the cable to tail uphill, I chose one downstream of the islands, for they lay in the equatorial current, which flowed here at around 6 knots. Letting go, hopefully stopped over the ground and judging that we were not making way ahead yet having a bow wave of considerable size as we stemmed the current, was somewhat unnerving.

These rocks were unique and rarely visited, a place to be set foot upon; but with such a torrent of water speeding past us and an unknown holding ground, I let Stuart go ashore initially with the gravimeter, whilst I remained on the bridge to monitor our position. The first boat to return to the ship had wildly alarming and warning tales to tell the next wave of Fids and crew eager to land. The waters teemed with sharks in incredibly sinister numbers slicing through the small wave tops around the boat. Worse was that these predators of the sea were complemented ashore by giant crabs, their quantity so vast that they completely covered the otherwise barren rocks. Stuart returned, and as I left the ship he warned me of

---

65 Sixty nautical miles.

the difficulty of jumping ashore. We were used to landing on icy rocks in heavy weather from inflatables, but this was different, as I was about to find out for myself. Weaving amongst the outer rocks towards the main outcrop with the sharks encircling us was chilling, but preparing to jump with the swell onto a mass of crabs, fearing not being nipped but squashing them and slipping on their slimy remains into the surrounding deep water full of sharks, was far worse. Remarkably as one's foot descended onto the crab-covered rock face, they instinctively raced out of the way, leaving a clear space, yet this was very unnerving as the crabs totally covered the spot one intended to step onto.

Once ashore, I climbed to the top of the main outcrop to visit the cairn built by the rare visiting sailors. Amongst the few bottles containing messages was one, which we left undisturbed, from Sir Ernest Shackleton, who had landed here from the *Quest* in 1922 on his final Antarctic voyage, when he died at South Georgia. Having traced his footsteps many times down south, it seemed very fitting for us to visit this little-known place of history associated with him in the equatorial mid-Atlantic.

Blue water

One of the mysteries, rather than problems, during our first voyage had been that of blue domestic water. Throughout that voyage we had fresh water so blue that beneath the taps and under showers the ceramics or tiles were stained a deep copper blue. Our distillation plant was made from copper pipes, so that seemed the obvious source of our problem. During the first refit we changed all the copper pipes to ones of stainless steel. But during this second voyage we still experienced exactly the same blue water. We had samples sent home and it was discovered that we had many hundred times more copper in our water than the norm, but we remained at a loss as to how this came about.

By chance, we met the US Coastguard icebreaker *Westwind* at Arthur Harbour, Anvers Island. The wind was blowing 60 knots as she lay there to two anchors. Her captain was very happy for us to lie alongside if I first laid out our own two anchors upwind as well. It sounds simple, but with the snow drifting blindingly off the nearby ice cliff in that wind, and the bow of an important ship under my lee to become impaled onto, the task was daunting. Laying the anchors across the wind just ahead of her was tricky, but fortunately the manoeuvre went well. I remember our meeting very well, for the large Yankee captain was quite impressed, and came aboard for dinner. We had a jolly evening, with the wind howling outside. The stewards were overheard to refer to the two of us as David and Goliath – I at 5 feet 6 inches and 11 stone, and he 6 feet 6 inches and 18 stone. During an interesting professional conversation he suddenly said, 'We've got darned blue water aboard the *Wind.*' As we related our experiences, finding common ground as to the practices aboard each ship and the actions taken to remedy it, the mystery began to unravel. It transpired that as we both sucked seawater into our desalination plants, krill and plankton would often block the intakes, and need to be blown clear as often as once a week. What we began to remember was that those creatures were very high in mineral content, and could possibly be the source of our copper. This was later proven. Seawater was being sucked through them as they lay caught against the filter, thus leeching their copper into our plant. It had been worth risking straddling *Westwind*'s bow to solve the mystery.

Nearly a black day in Black Channel

A new governor of the Falklands and the British Antarctic Territories, Mr Toby Lewis, arrived in Stanley during January 1971. He took passage with us around the peninsula early in this season. We had progressed south inside all the offshore islands on some either little-used or unknown routes, they being free of ice, the weather quiet, and my being in the mood and having the time for some exploration. It was rare to be able to go all the way from Deception Island inside all the islands to reach Crystal Sound, thence through the Gullet inside Adelaide Island, into the Laubeuf Fjord, and then exit into Marguerite Bay. I had managed parts of it at different times in the past but always been forced westward of the islets, rocks and shoals off Black Head, yet still inside the Biscoe Islands.

This time I closed Black Head, then Cape Evenson, both to port on the mainland, and shaped a course to pass through an unnamed channel barely two cables wide inside Marie Island. As we approached from the north it became evident that the tide was with us; this was not the best for a slow, controlled, exploratory run, whereas to stem the flow would have been far preferable, making steerage possible for a cautious advance over the ground. The channel had a small dog-leg, enough to prevent our seeing right through it. I had to maintain a good speed through the water to keep a flow past the rudder to maintain steerage, but with the tidal stream under us we were travelling far too fast to negotiate a narrow uncharted channel. Then the current appeared to quicken. There was no room to turn and abort. Going astern to attempt to arrest our way could have been disastrous, with our travelling broadside onto any danger encountered, so we hurtled on, with the only comforting thought that with such an inward flow of water, there must be an exit for it. As we turned out of the dog-leg, the reason for the torrent of water with us became apparent; a large iceberg was blocking most of the channel forcing the already strong tide to pour past it even faster on either side. The channel had been deep enough to allow the iceberg to enter, and I judged that it could not have arrived from the same direction as ourselves because of the depths we had experienced. I hoped it had found a navigable passage for us to use from the south.

But there was little time for such thoughts, for it lay right ahead. We frequently met bergs aground, and nearly always they fetched up one side against a rock or shoal, that side therefore foul whilst the other was usually clear. I made my choice and went hard to starboard to swing away from the berg, quickly realising that our room for manoeuvre was so tight that I immediately had to put the helm hard to port to stop the stern swinging on to it and to bring the bow away from the shoreline. I had put a helmsman on the wheel leaving myself free to roam, and I well remember shouting to him 'Go for your life!' as we made the two alterations. It was a beautiful, calm, sunny day, and many, including the governor, had been on the monkey island[66] to view the magnificent scenery. Even up there, however, they said that

---

66 The open top of the bridge.

they had felt the urge to run aft from the impending collision, as all did from the forecastle, where more were sightseeing. I, of course, was as worried as much about lack of water as about hitting the berg, but the least depth we found was ten fathoms. We entered Crystal Sound, where those who had ventured into it from the south had reported shoals galore, all described on the otherwise blank chart as 'position approximate'. There I once thundered astern for some breaking water only to realise that it was caused by frolicking seals. After that, as the sun shone, the water remained very deep, and I sat in the corner of the wheelhouse justifying to myself my action of going through what I then called Black Channel. Although inside passages, where currents keep them ice-free, were always good to know of, when outside the islands to seaward there could be dense pack to negotiate, I told myself inwardly that our latest exploration had not been one of my better pieces of judgement.

The next day we negotiated the mile-wide Gullet and then Tickle Channel, joining Hannusse Bay and Laubeuf Fjord to the north and south respectively. They were bordered by the high ice cliffs of Adelaide Island to the westward and the mainland to the east, and I had often been thwarted by heavy fast ice in the narrowest parts of these danger-free passages. The two large inlets frequently maintained fast ice, allowing mini ice shelves up to 40 feet thick to accumulate, barring their entire length. But that day the channel was open and a joy to navigate, and my chirpiness returned; I noticed, however, that the governor was still visibly shaken by the previous day's exploits. Progressing into the Laubeuf Fjord we came into one of the most scenic and romantic corners of the peninsula, with 7,000 to 8,000-foot peaks close to hand on both sides, and crevassed piedmonts fringed by enormous ice cliffs; and all about names of peaks and islands from Charcot's explorations, Pourquoi Pas Island, named after his ship, and members of his expedition being commemorated. Tanglefoot Peak finally elicited a smile from the governor, beneath which lies Lawrence Channel, later named after Stuart.

Attempted salvage

The *Lindblad Explorer* was the first purpose-built 'expedition' cruise ship to visit Antarctica. She was a little larger, at 240 feet in length, than the *John Biscoe,* and classed to carry just over 100 passengers. On her first voyage south in 1971, she ran aground in the Gerlache Strait, and the Chilean support vessel *Piloto Pardo* rescued her passengers and crew. The vessel was refloated and repaired, and sailed south again the following season, our second in *Bransfield.*

In the latter stages of this voyage we received an SOS from her, for she had become stranded at Plaza Point, one of the promontories between the cloverleaf of bays at the head of Admiralty Bay, King George Island, in the South Shetlands. Being fairly distant, we were unable to arrive at the scene until two days later, after everybody had got ashore to the nearby Chilean base. She was a sorry sight, having run ashore whilst steaming parallel to the coast. Admiralty Bay is a 12-mile south-facing inlet, mostly four miles in width; in a strong southerly gale after dark, her captain had chosen, somewhat extraordinarily, to steam across the wind within the bay, and not to stem the weather. Gradually losing his position to leeward, he had run inside reefs until fetching up with his bow so close to the beach that those aboard could jump ashore.

We anchored off our former but abandoned base, only a mile from the wreck, and boarded her from our inflatables. By this time a further gale had arisen, driving her against the shore. On any high tide she might have been driven further ashore, and we were not blessed with time, for our programme was demanding. Consequently action had to be taken immediately or the decision made to abandon her. I shifted to place both anchors upwind of her and dropped astern with all our cable out to fetch up as close as possible, then ran out our two strongest wire hawsers to her bow. In the meantime, a party led by the chief engineer, Tony Trotter, had boarded her with salvage pumps to begin to empty some of her spaces, and make a thorough inspection. The accommodation was a distressing sight, for every cabin was left with drawers and wardrobes flung open where passengers had attempted to grab some of their possessions before abandoning ship. But down below was much worse from our point of view; Tony and his team discovered that she had at least 32 visible ruptures in her hull. Some large gashes, others only where inlet pipes welded directly to the hull had fractured, but were, nevertheless, holes that had to be plugged. He assumed that more major damage must certainly have occurred to the ship's bottom, though at this stage not all the double-bottom tanks could be inspected. We worked on her for two days, and improving her freeboard, hauled her a very small distance offshore, but beyond one of the number of reefs which she had run over.

An important part of the operation was conducted via the teleprinter. As I handled the ship and the salvage party worked on the wreck, the radio officer kept a regular stream of advice coming from a lawyer in our London office, as I in turn advised him as to what stage of the operation we were at. Any attempt at salvage must be entirely for the benefit of the vessel salved, its cargo, and the owners of both. Any reward or benefit to the salvor, in this case our owners, myself and the crew, must be put out of mind. If at any stage the salvor worsens the position of the vessel being attempted to salve they can be held liable. Hence the constant stream of advice as we progressed, the lawyer judging our actions as if he were reviewing them subsequently in court.

We had a tough task on our hands. If we got her afloat my plan was to ground her safely elsewhere by strapping her alongside in the manner of a hip tow, where the stern of the towing vessel projects well aft of the stern of the vessel towed. This method, rather than towing her astern, was to improve control, to continue to pump her, and to prevent her from lolling, although that method would impede progress through any ice. If her condition was perilous, the destination I had in mind was a stone beach, rather steep but nearby, or if stable, to Deception Island, 90 miles away, where she could be beached comfortably on ash within the island's inner harbour.

The numerous inaccessible holes in her bottom plating denied us any such achievement, preventing us from lifting her sufficiently to float her over the outer set of reefs within which she was still imprisoned. Had we dragged her off we might have capsized her, or holed her further. So very reluctantly I put her back, her position then being only a little improved. Tony nearly wept, and all aboard felt deflated, their expectations dashed. Salvage spoils are divided approximately in the order of half the value salved to the salving ship's owners, in our case the Survey, and half to the crew. Of the latter half, the reward was allocated by rank, the captain getting a very large proportion. Tony had been planning the purchase of a country estate when plugging those holes, and he had actually got as far as getting some of the vessel's own engines and pumps running. A dream within grasp? Shortly after we had put her back, a tug, having taken two weeks to reach her from Cape Town, took nearly three weeks to pump her out with 14 portable turbo salvage pumps, of which we had just two. They plugged most holes and then filled her with polystyrene before towing her north. That dream was never really within our grasp.

Through the foothills of the Andes

At the completion of this season's work south, there were a number of reasons that made it sensible for me to get home in advance of the ship. To this end I took her to Ushuaia, intending to pick up a flight, only to discover that with the children of Argentinean service personnel returning to school from their holidays, all flights to Buenos Aires were fully booked for two weeks. We held a cocktail party on the evening of our arrival; I invited the military governor of the province, for our relations with the new Argentine government had undergone an improvement since my previous visit and I could further that by such a gesture. When I told him of my plight, he offered me a seat in his personal light aircraft, for it was returning without passengers to Buenos Aires the next day to be serviced. At 0530 hours the following morning, having handed over command to Stuart, I helped push the aircraft onto the tarmac, and away we went. With dawn breaking on a clear day, the mountaintops catching the first rays of sun, and the colours of the Andes foothills contrasting with the higher glaciers and the long, dark shadows of morning, we climbed through the passes, enjoying a magnificent spectacle. Suddenly we let down and landed at an isolated farmstead tucked in a deep valley. On the grass runway the pilot threw open the door and jumped out. After much embracing, aboard came a good-looking young lady who sat herself on the floor in the back, behind my co-pilot's seat amongst some baggage, for there were no other seats. 'You must sit up front,' I said, to which she demurred, insisting that she would be comfortable. As we cleared the mountains and headed to follow the coast and across the Patagonian plains, the pilot handed me the controls and a flight plan with some waypoints on a scrap of paper saying, 'Keep her at 6,000 feet.' At which he left me alone for the comfort of the rear fuselage, adding that he would have a rest.

He had greeted me that morning as 'Captain', and in our short conversation between taking off and landing again we had talked of the hazards of flying in Antarctica. Whether his understanding of what I did for a living was incorrect, or he just thought that I should be able to cope I shall never know. After about a three-hour flight as we neared Buenos Aires, he appeared, looking none too rested, and took over the controls. He was destined for the domestic airport, but realising that I was hoping to book an onward flight to Europe, he discussed the matter with the authorities and, finding out that the change could be made, landed instead at the international airport. After further discussions with the control tower he taxied to a waiting aircraft. I carried my bag across the tarmac and up the steps of a Swissair jet, the door was closed behind me, and we became airborne. What it was to have friends in high places!

11 RRS *Bransfield*: Third Voyage

Surprise visit to Mar del Plata

My next voyage in the *Bransfield* was once again not a full one, in that I flew to Montevideo to take command of her on the outward passage at the commencement of the season. This was brought about by the annual refit requiring very much more attention than expected, and therefore my being unable to take any leave at that time; the shipyard had become more disciplined in their control of work and we, in turn, had to ensure that every job was carried out completely as agreed. Taking leave after she had sailed, and that slightly later than normal, by the time I could depart for England winter in Europe had arrived. Leaving Southampton by train I met a Board member of Trinity House, who persuaded me to visit and be interviewed for a position on the Board after the sudden death of one of their number. He said although I might not want the position at that time it would stand me in good stead were I ever to think of applying in the future.

After a lunch and some individual interviews by the Board members available that day, as was then the custom, I went to Heathrow to meet Sir Vivian Fuchs and Derek Gipps, who were also flying out to join *Bransfield*. Fog was causing chaos at Heathrow and we spent over 24 hours waiting to depart. When we did, we were re-routed via Zurich and Buenos Aires. At Zurich we waited for ten hours, during which the three of us went into the city for some sightseeing; it was bitterly cold, and Fuchs had no overcoat or scarf to wear. Landing in Buenos Aires we had to await an onward flight to Montevideo the following day, so we stayed at the embassy. Here, during the evening, Fuchs swam for far too long in the pool. Flying again the following morning, the agents held us back from boarding in the VIP lounge, then boarded us after all the other passengers, only for there to be no seats left. This resulted in our having to sit on the floor of the aircraft for an hour, in a wicked draught.

On arrival in Montevideo I took command of the *Bransfield* and we sailed for Port Stanley. As we were passing the eastern bulge of Argentina, the weather deteriorated badly as did the health of 'Bunny' Fuchs, as all knew him. We found ourselves plugging into a full southerly gale, whilst the two doctors we were carrying became very concerned as to the condition of our director. Suspecting pneumonia, they requested that I get him to hospital. I altered course for Mar del Plata, which lay due west from us, but which meant that we were then broadside onto the weather, particularly the large swell, in which we rolled violently. Fuchs had by now become very weak, and had to be bound into his bunk, for fear of being thrown out. After an anxious night, with the doctors continually at his side, we made port and he was admitted to hospital. Derek and I briefly saw him the following day, none too well, but when visiting him the next, to advise him of my intended departure without him, we found him not only better, but regaling an admiring, mostly female, audience with tales of the south. Such was his stamina and charisma.

Sir Vivian was a slight enigma, in that while he was very fond of 'the ladies' he was steadfastly against them working in the southern polar regions, believing that to be a man's domain. He was a splendid boss, very easy to get along with, supportive, and a believer that once authority had been delegated there should be no further interference, so allowed his subordinates complete freedom to make their own decisions. Most outstanding for myself was that I left annually for between a seven- and nine-month voyage, with 'sailing' orders from him of one short paragraph, essentially 'Go south, relieve the bases, support scientific work, and carry out hydrographic survey when possible. Return home safely. God speed'. Of course itineraries were prepared in conjunction with the headquarters staff, and loading and personnel arrangements planned according to the locations of each base. But Fuchs in his sailing orders gave us complete freedom to arrange our activities in the manner we thought best, having regard to all the factors pertaining when south, particularly weather and ice.

Caught amongst the rocks

Landing Fuchs in Stanley to recuperate, and better in touch there with affairs of the Survey than in the Argentine or back home, we progressed south to the peninsula before embarking for Halley Bay. There had been no more surprises until en route we reached South Georgia. Here I had been asked to support a party investigating the southern side of the island and the nearby Annenkov Islands. It was a relatively calm day as I approached the shoreline near a large glacier when soundings showed the seabed rising suddenly and alarmingly. To add to my difficulties the outflow from the glacier contained a great amount of flour,[67] making the water opaque, as dense as milk. There was no breaking water, no grounded ice, indeed none of the helpful indications of where or not good or foul water lay. Neither was there any apparent relationship between the shoals I was amongst and any features ashore, and no hope of sighting the bottom through the murk. Every turn I made seemed to worsen the situation, and with not a single islet or above-surface feature to keep track of my movements, we spent several hair-raising hours extricating ourselves. In surprisingly pleasant weather for South Georgia the incident, although short-lived, provided me with one of the most insoluble circumstances to deal with. Ultimately patience was rewarded and shallow gullies leading to deeper water were found. We then ran for an anchorage off Bird Island, arriving just before nightfall; this is an awe-inspiring, dangerous-looking place, with a swell breaking on rocks close to hand, but on that evening it was a welcoming haven.

South Georgia

South Georgia has been neglected in my narrative so far, although we visited it frequently, my

---

67 Scoured rock suspended by water, produced as a glacier grinds over the rock beneath it.

first time there being in the *Shackleton* in 1957. It is because I knew and understood the island so much better in my latter years in command that I write of it in this context.

Captain Cook named the island at the time of his landing in Possession Bay, when he claimed it for the Crown in January 1775. It had been sighted, but apparently not landed upon, 100 years before, by a London merchant, Antonio de la Roche, and possibly sighted in 1502 by Amerigo Vespucci.

One hundred miles in length, it lies between 54 and 55 degrees south, less than 200 miles in terms of latitude further south than the Falkland Islands, yet for two reasons the difference in their appearance and climate is immense. South Georgia lies as far to the south of that important meeting of the warm and cold waters, the Antarctic Convergence or Polar Front, as do the Falklands to the north. South Georgia is also much higher than its north-westerly neighbour, being part of the Scotia Arc chain of summits, whereas the Falklands are not. That island group is comparatively low, damp and benign, and without permanent snow, whereas South Georgia is high, glaciated, cold and severe – a jewel set in a tormented sea; the island is often likened to Switzerland rising out of the Southern Ocean. Snow-clad mountains rise to a height of nearly 10,000 feet, set amongst razor-sharp rock peaks and pinnacles with long, broad glaciers descending dramatically from on high, and ice falls and crevasses that end abruptly with fractured, tortured, and calving snouts at the head of lengthy, deep fjords. There is permanent ice and snow on all high ground, except on vertical rock faces. In summer up to about 1,300 feet above sea level, pale green tussock grass, other green and tan grasses and moss, and rock, richly coloured with lichen, all interspersed with streams and pools, clothe the slopes down to the shoreline. This coast, wild and savage other than at the head of glacially formed bays, is sometimes ringed to as much as a half mile offshore with long strands and entangled masses of shimmering brown kelp. In sunshine and calm, the whole scene is a truly magnificent sight, and on an overcast day with a gale of wind and snow, it can be a foreboding, overpowering and awesome place. One could also visit it for a week and leave, having for fog seen nothing but a daunting shoreline of fierce rocks and kelp.

The magnificent scenery of Georgia is matched by the interest of its changeable and often violent weather, and the extravagance of its plentiful wildlife. Yet during the years of my initial visits to its principal harbours, this raw beauty was mildly scarred in these small locations by the whaling that took place there; so sickening was this activity that it greatly spoilt the overall impression of the island. As we steamed into the bays of Grytviken, Leith, Husvik and Prince Olaf, the first sight was of volumes of steam rising from the masses of machinery amidst which there were flocks of giant petrels, skuas and gulls wheeling and diving to gorge themselves on the discarded offal. Then the noise of the winches hauling up the dead whales and tearing off their blubber was heard ringing out across the water which, as we eased closer, became brighter and brighter crimson from the blood spewing off the flensing plan.[68] As we drew still nearer, the stench as the bellies were opened and the guts poured out was nauseating. On an early visit we naturally went ashore to observe the operation – but only once, after seeing the most perfect foetus, about two feet in length, taken from within the carcass of its mother

---

68 The boarded expanse up which the whales were hauled to be stripped (flensed) of their blubber and dismembered.

and cast aside. Around the harbours, the little single cylinder tug boats put-putting busily, working between the shore station and the whale catchers lying off, were the only cheerful thing about the scene, yet they too were involved in the slaughter, bringing bloated carcasses to the flensing plan from where they had been left on mooring buoys by the catchers. Around the coast the catchers made a fine sight steaming at full speed with spume flying from their flared bows and smoke streaming from their funnels. If one could forget the business they were about, they were welcome fellow voyagers in those waters, which was proven when one, the *Southern Lilly*, attended the 'holed' *Shackleton*.

Whaling was introduced onto the island in 1904, and several stations developed in the best harbours from then onward. Between 1946 and 1964, some 3,000 people, mostly Norwegians, were involved at these sites during the height of the season. Captain Cook had made known to the world the abundance of seal on the island, and after their consequent decimation and the establishment of whaling factories, an administration was put in place at Grytviken by the government in 1910, to issue licences for the taking of elephant seals and whales. This bleak outpost of civilisation, King Edward Point, with its bungalows on a flat spit of land and its own small wooden jetty, lay to starboard as one entered the harbour, removed from the whaling factory across the bay.

During the 1930s the International Whaling Convention began to control the quantities of whales that could be taken. Consequently pelagic fishing took over, and the factory stations of South Georgia went into decline. In 1950 it was decided that FIDS should provide a meteorological service for the whalers in return for the taxes they paid, which partly supported our operation at that time. A base of only five was established to reside alongside that of the government officials, although this was disbanded after just only two years, the Falkland Islands government taking responsibility for the meteorological station. In 1967, unforeseen by the government, whaling collapsed, and the administrative party was removed. BAS, having just erected a vast building to house them, took it over, and modified it to suit our requirements. At first it accommodated the expanding and successful biological programme running at Signy Island, later embracing other disciplines, whilst the base commander took on any administrative duties on behalf of the government, these being mainly associated with the growing number of visits by tourist ships.

As with the mountainous, ice-capped Antarctic Peninsula, South Georgia, lying in the path of many depressions, is subject to storm-force winds with regularity, but the polar terrain substantially boosts the ferocity of those winds, and makes them less predictable. When cyclonic winds line up with the mountain passes between peaks they become funnelled and increase in velocity. The air, colder at altitude over the interior, is heavy and thus falls rapidly to lower ground, accelerating as it does so. The consequence of these factors is at least a doubling of the wind speed by the time it reaches the shore in bays and fjords, or spills over the cliffs into the sea; yet whilst the passes funnel the strengthening wind, a mountain or buttress can create a wind shadow and shield one from it. This leads partly to the apparent unpredictability of the wind and to major variations of wind speed over small distances along the shoreline. I was once surveying in the ship with boats away, working only two miles distant around a headland, both of us in light airs, only to receive a call from them requesting to be picked up after a sudden increase in their wind to 40 knots whilst we remained in comparative calm. Before we could get to them it was gusting 60, entirely due to the changing alignment of an approaching depression. On one of the early voyages of the *Shackleton*, we once lay alongside the jetty at Stromness, the Salvesen Whaling Company repair facility, for the evening in reasonably calm weather. But before midnight we had parted all but one of our mooring lines one after the other, such was the ferocity of the wind, which eventually reached 90 knots; a lesson to have an even strain on all lines, which on that occasion we evidently did not. After a few moments lying to the one remaining bow line we also let go the anchor to a full stretch of nine shackles, and lay to both. Stones from the beach were being lifted by the gusting downdraughts of unimaginable velocity, carried out to sea and hurled against the ship; it was like being on the wrong end of machine gun fire.

A delightful and relatively harmless variation to the purely intense gust of wind is the williwaw, also to be experienced on account of similar terrain in the Magellan Straits. Caused by the irregularity of the terrain and sometimes eddies already present in the upper air, they are mini-cyclones often less than a few yards across, tight, fast-spinning winds that speed down the mountainsides to hit the water, stripping off its surface as they race across a bay.

In addition, cold downdraughts of great severity can occur unannounced, without a falling barometer and impending depression. Flying with some VIPs on a reconnaissance amongst the mountains in a helicopter from HMS *Endurance* in relatively quiet weather, we were caught in one as we approached the coast; such was its strength that we fell several thousand feet, the pilot all the while adjusting the pitch of the blades in an attempt to get a grip within the falling air. The urgent noise of the pitch changes, in addition to having our stomachs in our mouths until just a couple of hundred feet above the water the pilot succeeded in arresting the fall and landing us on the nearby ship, caused each one of us to collapse legless on deck as we disembarked. The governor of the dependency, two international observers and two ships' captains flat on their backs like jelly babies must have been a fine sight for the crew.

Better and safer conditions would normally be found out to sea rather than in anchorages on an inhospitable coast in very severe weather that can be exacerbated by the nearby terrain. Unfortunately, whilst this is generally true in most parts of the world, there are circumstances around South Georgia that negate that wisdom. The movement of the waters around its coast, splitting rather like the tides around the Isle of Wight, creates a building of seas where they meet. Particularly off the south coast, though elsewhere, too, this can create much higher than normal and extremely steep seas, without either being exactly wind-driven at the particular time of their occurrence or having any long-term direction. It is often rather like a gigantic lop. Shackleton almost certainly experienced and described such conditions when caught by a giant wave towards the end of his epic boat journey from Elephant Island to King Haakon Bay. In the *Bransfield* in about the same stretch of water, we fell off the side of such waves several times during the course of a few hours. It was only many years later that those of my deck officers who had been below at the time confided to me that they had thought that we would surely capsize. I fortunately could draw confidence from having observed the model of our vessel which when tank-tested returned to the upright having rolled to 95 degrees. Even so, anything approaching half that figure was both alarming and damaging.

Another phenomenon characteristic of this area is rogue waves; I have never subscribed to the term 'freak wave', always preferring 'rogue' implying an explanation, rather than 'freak' implying something extraordinary with little or none. Without such knowledge at the time, but now with satellite imagery of the ocean's surface and the mathematics of wave form better understood, which together explain how such waves develop into rogues, my choice does seem to have been the right one. A synchronisation of a number of waves appears to be the answer to the creation of these monsters. During my seasons in command I was to encounter two of these, both fortunately in daylight, enabling me to see them coming and, when already hove to, ready to take them head on at an appropriate speed; yet they still had a devastating impact.

Also of concern offshore at night is ice, in a latitude that has much more darkness during summer than offshore mainland Antarctica. South Georgia is not within the normal zone for pack ice, but does lie in the path of tabular bergs, some quite vast, and other shelf ice debris emerging from the Weddell Sea. They ground in their hundreds on the off-lying rocks and shoals, where they break up. This can suddenly produce a mass of smaller bergs, bergy bits and growlers which get carried around the coast and then northward out to sea. Thus a sea area free of ice on one day can be unpredictably cluttered the next. The many glaciers of the island, which calve extensively, also produce quantities of floating ice. Being glacial and therefore denser, and sometimes containing dark moraine, this ice sits dangerously low in the water and is difficult to see.

Nowadays the large flat-sided and flat-topped icebergs give a good radar echo, but a weathered berg, one which may have rolled over several times, the waves having smoothed its surface, may not. In my earlier years the radar was very unreliable, always needing to be tuned, making the location of ice a most uncertain matter. Geoffrey Hall, later Hydrographer of the Navy, spoke of an experience when approaching South Georgia in command of HMS *Owen* in 1961 to survey a part of the island's coastline. With the very occasional large berg showing on his radar earlier in the day, he later became uneasy in poor visibility, and hove to for the night. His sixth sense was probably an awareness of the colder, dank atmosphere created around a collection of bergs, for at dawn the weather cleared to reveal hundreds of bergs of all shapes and sizes surrounding him to the horizon, some very close; this was most likely the break-up of a monster berg. He passed through them for a further 200 miles while his radar operator, positioned below deck, managed to retune his set and report them! A micro-climate surrounding such a vast field of bergs is exactly what I would expect, and could be an invaluable aid to their detection.

Shag Rocks, an isolated outcrop, 90 miles to the west of South Georgia, was often surrounded by many grounded bergs. Such a micro-climate, together with the screeching of sea birds fishing in the surf breaking against them, was a valuable indication of this danger on passage between the Falklands and South Georgia, before the time of reliable radar. Wherever bergs are raised almost right ahead, they are best passed to windward, for their less visible calving debris will always drift away from them downwind.

Perhaps not surprisingly when considering the foregoing difficulties offshore, I never put to sea from South Georgia to escape the ferocity of wind, even in suspect anchorages, but preferred to remain tucked up near the land, even if it meant shifting anchorage to seek a wind shadow.

The wildlife of South Georgia is bountiful and beautiful, and I briefly mention the creatures to be seen in the order that they impact upon the approaching sailor. As with the Falklands, the albatross heads my list but here it is the wandering albatross, the largest of all flying seabirds. With their wingspan of up to ten feet, contrasting black, grey and white except for their long pale beak, and pairing for life, they can live for up to 50 years, but breed only every other. They circumnavigate the ocean on westerly winds, sometimes travelling as much as 250 miles a day. An elaborate dance ritual with outspread wings takes place when courting, forming a partnership and building a nest. This is usually a large moulded tussock lump rising to a foot or so above ground level. The offspring leave the nest after being fed for two years on fish and squid by the parent birds, then spend several years at sea before returning to their island of origin to breed themselves. They and the other three species of this magnificent bird are greatly endangered, particularly from the hooks of long liners.

The giant petrel is as unattractive and unpleasant as its cousin the albatross is endearing. It is not only an aggressive carrion bird, but when approached it disgorges its yellow, oily stomach content with velocity in defence. There are, however, many other smaller petrels – the white snow, the diving, and Wilson's storm petrel – all delicate and most attractive.

The skua, to be found on both the sub-Antarctic islands and the mainland coast, is another brute, which takes the eggs and chicks of penguins and other birds. Beautiful? As with the giant petrel, one has to admire the beauty of their form, flight and ability to survive the harsh conditions, even if their habits are not to our liking. In addition, shags, sheathbills, terns, a pipit, plus pintail and teal can readily be seen.

Of the penguins the king is truly majestic at some three feet in height. Its white breast sets off its orange/yellow lower throat, orange ear muffs and stripe on its beak, whilst its back and upper flippers are dark grey. Its gait is more like a monarch with gout as it waddles around its chosen flat terrain, with its single egg balanced on its feet, for it builds no nest. The chinstrap, gentoo and macaroni penguins are all much smaller, the latter with its singular yellow comb being the island's most prolific species. These are to be found in rookeries, sometimes of thousands, where their awkward strutting on land, relieved by the occasional tobogganing, belies their speed and agility in the water. Despite their prowess there, they are taken in numbers by the leopard seal. With its snakelike head and sharp teeth it is by far the most vicious of the seals, but the most infrequently seen.

The island is one of the world's three main breeding places for elephant seal. They establish colonies of up to 50 at the water's edge, often making wallows within the tussock, with a dominant bull presiding over a harem of cows and some adolescents. The bulls are gross; up to three tons in weight, they have a large proboscis that they regularly inflate, and they make a loud gurgling burp of foul breath when angry, from an often sore, red and gaping toothy mouth. They are usually scarred from fights amongst themselves or inflicted by leopard seals or killer whales. They have a multitude of female partners and during the breeding season they lumber through their harem to ward off any intruding adult male, sometimes squashing pups in their path. The young bulls continually stage mock fights amongst themselves in the surf, rearing up on their flippers and snapping at each other's throats. They were slaughtered almost to extinction for their blubber oil until the early 1950s after which a quota system for their killing assisted their regeneration. By the mid-sixties it was deemed unprofitable to take the small numbers available, so that they are now once again seen in great numbers.

The fur seal was likewise slaughtered for its fur and blubber. On my earliest voyages we searched for some without success, but on a slightly later voyage visiting Bird Island, off the western end of the main island, we found three, and reported the news with excitement. Their modern protection has resulted in them becoming so prolific that they are now regarded as a menace; such are the benefits of well-managed conservation. They have beautiful coats, are be-whiskered and have little ears; they arch themselves up on their flippers to move on the beaches and amidst the rocks and tussock of their nurseries at great speed. They are very aggressive and give a nasty bite. And we saved the little blighters from extinction!

Whales live their lives entirely at sea, and come to Antarctica to feed in the rich waters during the spring, summer and autumn, having both mated and given birth during their time in the more hospitable temperate zones. They are usually sighted when either breaking the surface to breathe, or breaching, when they leap out of the water to descend again with a tremendous splash, blowing, or slapping their tails onto the water's surface before sounding.[69] Their size and weight can rarely be appreciated from aboard. The largest, the blue whale, grows to 30 metres in length and 150 tons in weight. The seven other species to be found in these waters range in size down to the killer whale, which when fully grown is 9 metres in length and weighs 8 tons. Black and white, with a slicing dorsal fin, they are usually sighted in pods, or when attacking seals or penguins on ice floes in a group action that causes a wash of water over a floe to dislodge their prey and catch it, or when snouting their forebodies out of the water to locate their prey.

A further catastrophic overkill of these wonderful, enormous, and mysterious creatures resulted in us seeing fewer and fewer whales on each voyage, except for the killer whale, both less hunted because of its smaller size, and more visible to us in its predatory manner. On my early voyages a mother suckling her calf, or swimming closely together, were often seen. Once at anchor in the *John Biscoe* a blue whale, half the ship's full body length in size, therefore about 30 metres long, accompanied by her calf, scratched its back against our hull for almost half a day. Unimaginable proximity to nature.

The collapse of an ice cliff

In January 1973, striking south from South Georgia for our annual visit to Halley Bay, we left Cape Disappointment, the island's eastern extremity, astern and commenced the circuitous route into the Weddell Sea. Halley Bay Station was renamed Halley Research Station in 1977

---

69 Diving to great depths.

after the bay had completely disappeared, but there were still deep fissures in the ice cliff to be found. These were not too distant from the base site, and we knew them as creeks we could berth in. Our visit was to be of some weeks, for we carried the annual relief stores, plus materials to rebuild Halley III, which we were to assist with. The remaining fast ice, together with some of the snow ramp, gradually broke away, causing us to shift ship several times, first to the lower remains of the ramp, then alongside the cliff itself, in a position where the cliff top was at bridge level. It was only possible to work cargo by lifting it high above the well deck then just a few feet inland onto the ice shelf, to be hauled away quickly from the edge. Because of our protracted stay and the existence of hairline cracks on the shelf ice surface alongside our 'berth', we had taken the precaution of fixing some light cordage across them, to help us identify any movement and likely calving. We were working two 12-hour shifts consisting of base members, the ship's Fids, 12 Royal Marines who had volunteered from the garrison at Stanley to help with the rebuild, and as many of the crew as could be spared.

One evening immediately prior to the 2000 hours shift changeover, the adjacent high ice cliff calved without warning. Those going ashore would normally assemble with the incoming team on the top of the ice cliff, to exchange information and banter regarding the building work, before boarding the vehicles for the journey up to the base site. Approximately 300 yards of cliff face for some 30 feet landward, beyond our telltale markers, and 50 feet out of the water, sheared off, and the ice alongside the vessel fell onto her. I was going up the companionway to the bridge at the time to take a general look around, but particularly to observe the handover between shifts. I was thrown off my feet but got to the bridge to see a multitude of ice blocks, many the size of houses, covering the decks.

My greatest concern was that nearly 100 men could have been on the cliff and thrown down into the water amongst the seething blocks of ice there. This did not seem to be the case, however, because I could see none there. Another worry was that half that number would have been on deck, waiting to disembark, or just arrived back from the base, and could be buried beneath the ice. More men than I expected appeared, to axe, shovel, push and roll the ice off the decks as I, fearing a major catastrophe, put in hand a head count both aboard and at the base. Fortunately all but one man was accounted for, and he was later discovered behind the door of a deck store, barricaded in by ice. The day shift had arrived and been relieved early. No one had been on deck, and all bar that one had been either up at the base or in the ship's mess rooms.

When the cliff calved, the weight of ice hitting primarily one side of the decks was such that our main deck, with over 12 foot of freeboard at the time, rolled beneath the sea surface. The resultant immediate counter-roll was beneficial in that it instantly shed a great deal of the larger lumps, but also began to expose flattened rails, ladders and ventilators, and distorted hatch coamings.[70] Fortunately, as we initially rolled towards the ice shelf, the explosion of ice from the shelf face at hull level had forced the vessel away from the cliff, so that the masts, particularly the main, carrying the crow's nest, and cranes did not touch the cliff. As I reached the bridge with the ship still reeling from side to side from the impact, I became aware of a

---

70 The raised side-walls protecting openings in the deck such as entrances to cargo spaces.

further piece of luck. The collapse of ice to our nearside was setting off another fall on the opposite side of the creek; the enormous conglomeration of ice emanating from that second cliff fall further filled the creek, preventing the vessel being sent flying across it by the nearside fall and smashing broadside on to that opposing ice face, almost certainly becoming seriously damaged, and possibly holed around the waterline.

As the two cliffs calved, enormous chunks of ice rose from deep below, surging out of the water, through the existing loose ice, then heaved up and down before settling in equilibrium. Consequently my next fear, after the possible loss of so many men, was that the underside of the hull would have been holed, and both rudder and propeller very badly damaged or carried away. No sign of any rupture was apparent, although most tank sounding pipe caps on deck could not be accessed, buried as they were by ice. I had the chief engineer start the main engines, and eased forward the lever controlling the variable pitch propeller. The vessel immediately responded by inching ahead, indicating that we did still have a propeller. I then gently put the helm over, first one way then the other, and our fine, strong, safe ship, responded with a wriggle of her stern. After some further checks, and receipt of knowledge that all personnel were safe, I took her out to sea into the quiet of some surrounding pack ice, rather like a bedraggled dog shaking itself on emerging from a pond. It had been a near-miracle that that nothing of consequence had happened to men or ship.

Early the following morning, to enable the 0800 hours shift to get away, I took the ship back alongside the pristine, newly exposed high ice face, believing, or perhaps just hoping, that it was then more likely to be stable than the rest. To our relief, it did not calve again during our stay.

A dog wins over the ozone hole

To end the voyage, we made our usual pleasant voyage through the tropics to a rapturous welcome in Southampton, marred only by some irritating nonsense in the press; the presence of the ozone hole was being discovered at Halley Bay and the Argentine Islands, and the Survey wished to utilise the occasion of our arrival to release a statement concerning the work. However, we had brought home with us an arthritic husky which was to go to someone's home to see out its days in luxury as a pet. The press became aware of this and blew it out of all proportion in articles condemning such an action as irresponsible and a waste of money, totally eliminating the first hints of a major discovery.

12. RRS *Bransfield*: Fourth Voyage

This voyage of the *Bransfield* was rather different from the previous ones in more respects than one. On 10 October 1973 we sailed for the United States with my wife, Ella, aboard. I had been lucky enough to meet her in Edinburgh during the building of the *Bransfield* – indeed she had been the first person on the ship's payroll! Being a secretary, she helped me in the shipyard at weekends by assisting with the ship's paperwork. However her growing interest in the vessel resulted in her also accompanying me on visits to the ship, whereby she knew more detail of the initial construction than most who sailed on her. Three years after our marriage in Winchester, our director Sir Vivian Fuchs agreed, contrary to his normal ethos, that her presence aboard would be beneficial to me and therefore to all on board.

An agreement between BAS and the American National Science Foundation (NSF) took us to Rhode Island to embark stores for their Palmer Station, previously our Anvers Island base. In return we were to be supplied with fuel, which was then in short supply worldwide, in North and South America. On passage we hove to in order to ride out the tail end of a hurricane off Newfoundland, and on the radar we saw many ships similarly stopped, marking the passing front. This was strangely different from seeing only the echoes of innumerable icebergs in our lonely existence down south. It also gave Ella her first taste of bad weather, during which she coped well, never missing a meal, in fact enjoying the spectacle, which augured well for the rest of the voyage. On our arrival at Rhode Island, the US Immigration Authorities wished us to dump most of our fresh stores, under their import regulations. This I refused to do, calmly stating that I would leave without berthing if they insisted on that happening, not giving any thought either to our agreement, or that we carried very little fuel; fortunately the NSF soon stepped in and we kept our eggs and fruit, and embarked their stores. From there we went to Norfolk Naval Base, Virginia, where we refuelled to capacity. Then striking southwards for Montevideo we made a scenic run through the West Indies, often closing the islands. For all aboard it felt rather like a cruise, and the crew, mindful that I had my wife aboard, were overheard to say that the 'old man' was, after all, human.

During the passage south I heard by cable that for the second time in a few years a Trinity House Board member had died and that the vacancy was to be advertised. Being somewhat half-hearted about leaving my present job, I applied, but being unable to attend for interview until the following June, and believing that they were unlikely to consider me for a number of reasons, particularly the length of time before my return, I proceeded to forget all about the matter and get on with the voyage.

Exposure of a different kind for a film crew

At Montevideo a television producer, Ned Kelly, and a journalist, Anthony Smith, came aboard, together with their camera and sound crew. They were to make a film, *Wilderness*, for BBC Bristol, and Anthony was to write the accompanying book. Ned and Anthony joined my wife and me for dinner in my cabin on their first evening aboard, and it quickly became apparent that they wished to target pollution and disturbance to the natural environment by all those who had research stations in Antarctica, rather than extolling the virtues of the work we did and the magnificence of the place and its wildlife. A slightly uncomfortable evening ensued, but without any falling out; I downplayed the pollution that occurred, and enthused about the cost-effective benefits to science to which the research work of the Survey contributed. I stressed the evidence that they would see of that, plus the unimaginable extravagance of the scenery and incomparable wildlife they were about to be introduced to. That seed was well sown that night, and after a few days around the bases they altered the outline of their programme to reflect our work and the inspirational surroundings it took place in.

They became pleasant companions, involving members of the ship's company as they went assiduously about their tasks wherever we went. When we were anchored off Palmer Station at Arthur Harbour to land a few Americans who had also embarked at Montevideo and deliver their stores, we landed the film crew on Torgersen Island, where there was a large penguin colony. The day was pleasantly fine and calm when they set off for their destination, barely two miles distant. But by noon the wind had gradually increased to a full gale with driving snow, and the temperature had dropped. It became obvious to us that even the most ambitious cameraman would not find it possible to work, so a normal pickup was set in motion to retrieve the party. Normal, because two things made extreme urgency unnecessary; first, they had with them an experienced Fid whom we had expected to ensure that they were properly clothed and equipped. Secondly, parties including cameramen nearly always wished to push the boundaries. We amateur photographers put our cameras away as the weather deteriorates, but they always wish to record it and, with their high quality photographic and sound equipment, are able to do so; the raw savagery of nature at the height of a storm and the penguins' ability to confront it would be manna to them. In fact they turned out to have been in shirtsleeves, and in a very short time we nearly lost the whole party from hypothermia, for even when embarked by launch they had to endure the return passage to the ship, where fortunately, after some care, they recovered. Happily they went on to enjoy the remainder of their venture with us and produced a good polar episode to their film series, and Anthony a good book.

Sojourn in the Beagle Channel

Prior to going to Halley Bay we went north from the peninsula bases to visit Punta Arenas to pick up personnel, particularly the new director, Richard Laws, a zoologist, mentioned earlier as being a Fid at Signy Island, back in 1948. Unusually, we managed to convince the Chilean authorities that a passage through the Beagle Channel was acceptable, rather than an approach to Punta via the Straits of Magellan and through their eastern entrance. The navigation was intricate and involved but mostly in deep channels, so of no particular concern, but the route was of great beauty and interest. I have an abiding memory of both our pause at, and our departure from, the preliminary anchorage we took in a small cove near Puerto Williams, to await dawn for a daylight passage. Its association with Fitzroy, who had lain there several times in the *Beagle* and written of it in his journal, was fascinating, as were the striking differences from the Antarctic Peninsula. Here there were glaciers, but green trees and shrubs fringed them. Birds were abundant, but ducks and land species rather than maritime, and replacing the chill of down south, the temperatures, just above freezing, were positively balmy.

Some Fids and crew landed and brought back flowers – as had Darwin – but on our visit they were collected for my wife. It was New Year's Eve, and as I anchored at 2100 hours, which coincided with midnight at home, Eric the chief steward had champagne brought to the bridge where he also had Scottish dance music playing. So Ella was able, as was her custom, to wish her parents a happy new year at home in Scotland from wherever she was, albeit out of their hearing. At midnight almost the entire ship's company was gathered on the helicopter deck, its blue and orange landing lights around the perimeter adding to the party atmosphere. As 16 bells were struck to ring in the New Year, the flowers, a bottle of Scotch, a piece of coal and a black bun were presented to her as we all linked arms to sing 'Auld Lang Syne', underlining the camaraderie that existed between all those aboard and their growing fondness for my wife. Then only a few hours later in half-light as the sun rose unseen behind the mountains, and with a williwaw or two racing across the bay, we weighed anchor. As we left the cove that had been sounded out by Fitzroy, we ran towards a shoal central to the entrance, which I can only imagine would have been missed by him tacking on arrival and departure and hugging the steep shore to windward, so never passing midway between the headlands.

Concern regarding safety of personnel

From the point of view of the ship and her welfare, all went well during the major part of this voyage, although we seemed to have a succession of concerns regarding the safety of individuals, starting with the episode at Arthur Harbour. Calling at South Georgia, we met the Italian Antarctic Expedition in the extraordinary form of a lateen-rigged yacht, the *Sangiuseppe Due*, owned and under the command of Captain Catt, a tomato millionaire from the Yacht Club of Rome! I feared that if he ventured much further south he would become a liability, but am not sure whether he ever did. He rather irritated me, during our stay in Grytviken when two of our mess boys got lost one afternoon on a walk in the hills behind the whaling station. Darkness fell and with no sign of their return we established a search party. Unfortunately Captain Catt was aboard at the time and proceeded to attempt to take charge, getting both into a flap and in the way. Perhaps rudely, we told him to shove off. The mess boys turned up very shortly afterwards; they, unlike the film crew on Torgersen Island, were properly dressed and equipped, but they had told no one when or where they were going – the second cardinal sin.

When we reached Halley Bay the ice cliff gave us only a little trouble from calving, necessitating shifting ship and re-mooring. There seemed to be more penguins than usual on the ice shelf and they, like Captain Catt, seemed to want to be involved with our operations, particularly the digging of the deadmen for the mooring lines. When not actually in the way they would line up, nodding their heads and observing the work as if saying amongst themselves 'They did it better last year'. The weather was not as settled during our visit that season, and although one very strong gale did not affect the ship, lying low beneath the ice cliff, it did cause all interchange with the base to cease. The two-mile track up to it was marked by flags and oil drums for the men to pass safely through the crevassed areas, but in the blizzard conditions we were forced to collect everybody from the route, picking them up on sledges behind one vehicle and returning them to the ship as the storm broke. A head count at both the base and ship showed that one person was missing. The base commander. Concern grew as darkness fell, but little could be done overnight to find and retrieve him without endangering others in the white-out.[71] We just hoped that his knowledge and experience would assist him, which they did; the following day, when improvement came, he was sighted emerging from a 40-gallon drum, which he had dug deep into the snow and sheltered within.

Once more we ventured further south to visit Shackleton Base, to remove the build-up of snow at the entrance to its access shaft in order to enhance its use as a depot. On our passage back to Halley Bay we were asked to lift and return a survey party that was on an extended journey from the base. It was late evening when we made contact with the party, and I commenced searching along the cliff face of the shelf ice in their vicinity to find the best place for them to embark. There was a strong offshore wind but a heavy onshore swell, enough to prohibit our attempting to lie alongside. It was difficult to find a clifftop exactly the height of the prow to enable them to board over the bow, and when one was found it had a spectacularly curled cornice, quite unsafe for them to venture across. I chipped away some of this and in doing so realised that our rise and fall on the swell was much greater than I had previously thought. Using the ship as an ice axe was not unusual, but at every touch we also rode up and down several feet, the bow grinding against the cliff and the shuddering from the impacts running right through the ship, and this was unnerving. Were it momentary it could have been tolerated, but when the chipping of a V in the cliff was complete, and with the ship in position, it took hours for the party plus vehicles, sledges and stores to be embarked in the dark during the moments when forecastle and cliff were level with each other. Each man, and every valuable piece of equipment, was attached both ashore and aboard to a safety line; such extreme caution was necessary before even an advance to the cliff edge could be made, and steps taken to come aboard. It was only at dawn that the operation was completed. Had I realised that it was to be so difficult I would have waited for daylight before commencing, yet waiting around in the foot of the Weddell Sea at the end of the season was not the best option, either. With the constant hammering and shuddering over several hours, plus the tension of getting each man safely aboard, the episode ranked as one of my most trying on that voyage.

The psychology of being master

Caring for those aboard was an important part of my role but was not rocket science. It was

---

71 An atmospheric condition in which daylight is diffused by multiple reflections between a snow surface and an overcast sky. Contrasts vanish, and it is impossible to distinguish the horizon or any other features.

mostly common sense based upon some definable humanitarian lines in recognition of our particular circumstances, probably none too different from that of any armed service commander, except that it was the elements, the unexplored and the isolation that were our constant enemies. I had to remember the young age of the crew and, apart from a few senior officers, their total inexperience of our lifestyle, the conditions aboard and the environment we worked in. I had to remember the effect upon their mental state of a nine-month voyage, particularly at the commencement when such a long haul still lay ahead, or when our circumstances or events at home brought that period away from civilisation into even sharper focus. I also had to remember that most young members of the crew were unlikely to have been away from home at all, and many of the older ones rarely for more than a couple of months; to be aware, too, of the effect at times that our remoteness, combined with fear of disaster in ice or uncharted waters, could have upon the individual. Mentally toughened by my own experiences and my knowledge of the ship's capability, it could have been all too easy to overlook that others were often frightened.

Once, as an exhilarating afternoon came to an end, working a totally virgin exit from Stonington Island, north through the Debenham Islands, twisting and turning amongst rocks and shoals, I turned and spoke to the second officer with me on the bridge. He was a pleasant and capable 26-year-old, and when I remarked what an exciting day we had enjoyed, seeking out a new route, I found him visibly shaking and just managing to reply, 'That was the most appalling day of my life'. Although he had been fully occupied tending the echo sounder, calling the depths, helping me watching for shoals, plotting where we were as best he could from significant features, and recording what we found, he had been terrified. I found it hard to understand, but had to remember that his was not an isolated case and that such fear could occur in others. Possibly easier to understand, but still worrying was the deck officer who, when we repeatedly fell off the side of monster seas near South Georgia, had not only been convinced that we were about to capsize, but had caused others to fear it too. They were down below in his cabin; outside it was pitch dark with a howling wind and ferocious cross sea. On occasions such as this I needed to remember that my own experience and ability gained was assurance to myself, but until a crew had sailed with me for a while and experienced difficult conditions, there was no comfort for them. So I had to be careful at times as to what I said on the bridge, often exuding overconfidence, for the opposite quickly trickled down below decks. Likewise I could never show despair, for it could be ruinous for morale. I well remember, during both my escape from within the anchorage at Signy Island, and the night of terror off the Biscoe Islands, trying not to give a hint of the desperate situation that we were in. Frustration on my part, usually somewhat exaggerated and with some expletives, was mostly taken as amusing, but I had to try not to show mood swings.

I had to remember, too, the effect of news, whether via receipt of mail or radio telegrams or on the wireless. Mail was the biggest enemy of morale. It rarely brought happiness, but mostly longing, sadness, even depression, and the lack of it brought despair, even jealousy and envy of those who were in receipt of it. My art was to know my crew, their personal circumstances and from whence they came, and to monitor them carefully when internal or external events dictated it. On a vessel with a much larger crew it would have been impossible for me to play the role I did. I knew all my crew by Christian name, but that never led to familiarity. I tried to be a father figure. I was akin to the head man of the village or tribe; although I was in charge of them, I was always quite close and concerned as to their wellbeing, caring for them and about them, though the majority were probably never aware of that. Considering them each individually helped me build an entirely happy crew. There is much spoken of the loneliness of command, but I rarely felt it. There was no loneliness as the head of a happy family. Yet I was a disciplinarian and never tried to be popular.

In hindsight it never occurred to me to ponder whether I was even liked. I attempted to favour no-one. To those that deserved praise I gave it them privately. I tried not to reprimand people publicly, though often in the immediacy of a situation, perhaps on the bridge, it was difficult not to do so. Some took it hard, and then some repairing had to be undertaken, for we all had to work closely together for months on end and grudges could not be held. After a reprimand was given there was usually some minor social occasion, aboard or ashore in Stanley or at a base, to exchange some pleasantries with the individual concerned, and the equilibrium to be re-established. There was little or no chance to replace any of the officers or crew if they were totally incompetent or misfits, although on this my last voyage I sent home an engineer, a sad alcoholic, from Punta Arenas. Also, on a previous voyage, one young deck officer was continually seasick, probably through some thyroid malfunction, rendering him incapable for most of the time, and he had to be landed without immediate replacement.

Conversely to the care of most aboard, Stuart thought that I expected too much from my wife. Ella was slight but very fit and tough. She coped extraordinarily well for someone who had never been to sea before. She was never seasick, in fact would watch the spectacle of bad weather from the bridge, where she also enjoyed being at the centre of things but knew when to retire if matters became fraught. She coped with climbing up and down rope ladders to take passage in the boats and inflatables, and from them onto bad landings ashore in all weather, though the bosun Robbie Peck, a stalwart Falkland Islander, appointed himself as her minder and was frequently at her elbow to assist. Once at Bird Island we prepared to go ashore and in the cabin put heavy boots on. She tripped over the door sill and stumbled about in hers, and I grumbled. The scientists ashore recommended better footwear and lent us some more suitable ones when it was discovered that I had given her some oversize boots of mine. Yet she had negotiated the Jacob's ladder on the slippery rungs without a murmur. Justice prevailed when it was I who slipped amongst the tussock grass and was bitten in the buttocks by a fur seal. The only tears she had were when steaming up Southampton Water, knowing that the voyage was over.

After Halley Bay we made a visit to Mar del Plata to acquire more fuel from the Americans, so this was truly a voyage with a difference. Not so good for the southern programme, but pleasant for us was that despite the arrangements and assurances no fuel was available on our arrival. Nor did it materialise for nearly a week, and this allowed everybody to enjoy the pleasures of this coastal holiday resort. Being cooler and less humid in midsummer than Buenos Aires, it was a favourite destination for those who could visit it.

Surprising news

Making the final tour of the season, we revisited South Georgia before going back to the peninsula. It was whilst we were berthed at Grytviken that I received the totally unexpected message that without further interview I had, subject to a medical, been elected to the Board of Trinity House. The purpose of having my wife sail with me, apart from our being together, was to let her experience all that the coastal regions of Antarctica had to offer, and to understand our work and what I would be undertaking on future voyages; as far as we were concerned it was a one-off event, for we wished to start a family, yet I expected to voyage south for several years more. Nevertheless, we celebrated with a bottle of cheap champagne from South America, as we stood on the wing of the bridge in sunshine. On our own during the afternoon, we climbed Mount Hodges at the head of the bay, to contemplate our future in the sharp clear air, as we absorbed the beautiful panorama, including the bay spread out way beneath us, harbouring the toylike *Bransfield*. The ship's officers watched through telescopes from the bridge and saw me helping my wife onto the steep summit, where she crouched down, finding the vertical drop beneath it too much for her. On our return to the ship, it was not she who was teased for the technique of her final assault, but I who was castigated for pushing her to her limits.

Rothera

Next, we made the last visit of the season to the bases on the peninsula. These were declining in number, the emphasis at this time beginning to be on fewer, larger scientific bases rather than a proliferation of smaller ones for topographical and geological surveys, the work on these disciplines having been exhausted and the political requirement for them having been made redundant by the signing of the Antarctic Treaty.

Following this premise, Rothera Point was again being considered. Within the Laubeuf Fjord, it had been thought a desirable location for a large base and location for the aircraft, yet the idea had been abandoned when the *John Biscoe* repeatedly failed to reach the site because of fast ice. The recent seasons of ice had not been too difficult, and in 1972 I had passed right through the Laubeuf Fjord and close by the intended site with little trouble. Approaching from the waters to the south of Adelaide Island, then north, I paused off Jenny Island to allow some scientists to land and look at the tiered, raised beaches to be found there which are of great geological significance. An ice sheet of much greater mass than that of the present one had existed during the last ice age, 20,000 years ago, weighing the continent down, but since then it had decayed into the ocean. The removal of such a weight results in the earth rebounding, glacial isostatic adjustment, clearly indicated by the raised beaches which help geologists understand the history of the Antarctic ice sheet.

We then moved northward to arrive off Rothera Point. Here I found an unexpected plethora of rocks and shoals off the headland, with, of course, an associated multitude of grounded bergs, making the finding of a safe anchorage within an acceptable distance from the shore very difficult. I spent some considerable and frustrating time twisting and turning, running through the shoals to find some pattern in them, and hopefully some space to anchor in. The terrain close by looked most promising as a base site, so I was determined to find at least a temporary anchorage to get ashore from, until the whole offshore area could be surveyed to establish a better permanent one. Finally, almost in desperation, I let go the anchor in the best spot I could find and we brought up. The previous hour or so had been fairly tense on the bridge with little said except for soundings being called out and orders given to the helmsman, whilst the rocks and shoals were carefully monitored. As we settled to a short scope of cable, the tension was broken by my wife appearing on the bridge saying, 'Poor Henry has died!' She had been immersed in a book about Henry V and was most upset at his demise, which nicely changed the mood; we all had a great laugh. Stuart then took the motorboat away to sound immediately around the ship and found a nasty shoal close under her stern, but we stayed where we were with just enough room to swing, carefully monitoring both her movement and the weather.

Once ashore we found that the area could not only provide an excellent base site but believed it could, with some work, provide a long, flat shingle runway. On return to the ship I wrote a glowing report on the site, playing down the difficulty of the anchorage in the hope that a better one would be found. The following year a base was established there. Meanwhile, on this first visit from seaward, and with the sure knowledge of a good future base site under our feet, we were able to turn our attention to one of the most glorious Antarctic panoramas, across the Laubeuf Fjord to Pourquois Pas and other islands, and then to the mainland, rising to 8,000 feet. The next season the ships discovered that although our approach to the site from the eastwards was as foul as we had found, they were able to approach it from the southward such that *Biscoe* and *Bransfield* were able to lie to Mediterranean Moors[72] with their sterns close enough in deep water to refuel the base by hosepipe. Shortly after, a quay was constructed in that location, and the ships now lie against it.

The grand finale

The season was over, done and dusted – or so we thought – as we left the fast-cooling, pristine southern continent, with its lessening hours of daylight, for warmer climes, but initially duller ones, until we could also leave the Falklands astern and head for the tropics and home. Modern thinking has it that equinox gales are a myth, but I must admit to an archaic view that those times of the year produce some howlers. Perhaps it is correct that the gales around an equinox are not directly related to it, but it is true that in the northern hemisphere the time of the autumnal equinox coincides with the end of the hurricane season; the tail end of some hurricanes get across the Atlantic either in their own right or, carried and enjoined with regular depressions on the jet stream, arrive on our shores in Britain whenever that high-speed flow of air brings them in our direction. Likewise there are circumstances in the southern hemisphere that bring storms across the Southern Ocean simultaneously with the equinoxes. So the old adage is none too far adrift.

---

72 Lying at a right angle to the quay, with an anchor or two ahead and stem lines to the shore astern.

This late March crossing of the Drake Passage was to prove that. We were to experience not just one storm but a whole family, relentlessly pursuing one another, as they were formed by the clashes of warm and cold air. It transpired that this was to be my final voyage down south and so it was also my final crossing of the Drake Passage, those changeable waters between Cape Horn and the South Shetlands. I had seen them calm, with or without fog; in ferociously shifting gales with unbelievably immense swells; and with a massive single tabular berg measuring 60 miles by 40 dominating a portion of the seas around us which, after its break-up, were strewn with hundreds of the resultant more minor bergs. But Neptune was to show me one more face of this ocean, and have one final onslaught upon one of my minuscule red ships, testing my abilities to the limit one last time. Extraordinarily, after 19 seasons south I was about to endure the worst storm of my career, with my wife alongside me to witness it.

The wind freshened from the north-west as the barometer fell shortly after our leaving the peninsula for the Falklands; not unusual. Nor was the increasing swell making for an uncomfortable motion as it pounded the port bow. Snow hindered visibility, making it difficult for us to spot the stray pieces of pack that had spewed from the Weddell Sea in the increased autumnal northerly drift of ice. This was caused by new ice forming in any open water between the older floes further south, as temperatures there dropped dramatically, inhibiting the floes from closing together, thus producing the spreading action which forces the older ice northwards. The temperatures, even at 60 degrees south were now decidedly colder too, and icing on the upper works was a consideration until we reached latitudes appreciably further north. Our progress in that direction was slow, as we were taking the gale on the port bow causing both pitching and rolling, though it was not necessary to heave to. We forged ahead until the barometer rose and a south-westerly came in somewhat stronger than its north-westerly forerunner – but, of course, colder. With the wind and sea now on the quarter, we picked up speed, but not comfortably for the swell had increased, now coming from due west, making us roll more violently but without the pitching. Patience, watchfulness and caution were required for one last time in these waters. The barometer would rise still further; the storm, as all do, would pass; the wind would settle to a nice breeze as the sun reappeared, and the chance of meeting ice would diminish. Of course the grey Southern Ocean would become unpleasant again when the ridge of high pressure following in the wake of the depression collapsed, but by then we would be away and clear of its ice and ferocity, or so I thought.

The barometer did rise but only to fall again immediately, this time more steeply and to a somewhat lower point than on the previous occasion. The wind, sea and swell rose, and were such that I was forced to heave to heading northwest. As it was the end of the season we were light ship, devoid of any cargo save empty oil drums and bags of mail and a few pieces of machinery for repair. She was a heavy, stable ship, but nevertheless in this ballasted state was distinctly lively compared with her fully laden motion, and was catching the wind on her high 'cock-a-snook' protective bow. Stuart and the chief had ensured that every conceivable empty tank was ballasted, which improved both motion and handling but made her marginally wetter; but I held little concern about solid water coming aboard, for she was sound and secure. But as the winds reached 100 knots it became a matter of making sure we came through it relatively unscathed; the scow and motorboat on the foredeck, secured as best as possible, would have to take their chances.

As the barometer rose again and the wind shifted marginally from north of west to south of it, and with the gusts rising in intensity, we sent a weather report to the Falklands. The remnants of the met station there, after the cessation of the service for whalers in the mid-sixties, sent us a note crossing with our report, warning us of depressions ahead! Without referring to the log, I lost track of the rises and falls of the barometer after the first 48 hours, concerned only with handling the vessel up and over each swell, a measured 70 feet in height, and possibly more after the recorder had malfunctioned in the chaotic maelstrom of the sea about us. There was frequently a wind of 120 knots screaming past the wires, aerials and masts, and through every aperture as we rode over the crests, yet there was almost no wind in the troughs. I increased the power as we rode up the mountainous faces so that we did not fall off and broach, alert to act further near the top of each breaking beast where the sea and swell became one and broke, deluging us with solid water. Then I eased the power as we sped down into the following trough, trying to avoid burying the forecastle at its foot before climbing the next mammoth.

During nearly every cycle, as each trough was reached, regardless of how I adjusted the power for the descent, the vessel was virtually in freefall down the face of the swell, so that she slammed, violently and loudly into the bottom of the trough, the reverberations running through the length of the ship as if she were a taut wire. She was strong throughout and very solidly built, with additional steelwork in the forefoot and concrete reinforcement between the stem and first watertight bulkhead. I had distorted the shoulder plating and frames behind it in ice, but rarely the very fore part, and never that or the entrance in a seaway. Yet thoughts as to how much of this she could take passed through my mind. What might fracture, which bit of equipment shatter, which bolts shear? I was mindful of the time when from shock in ice, bolts within the armature had sheared loose and destroyed the windings, rendering us without power. Not for long, though, did my thoughts stray, for I needed the utmost concentration to maintain as safe a control as possible – but I did manage to think of those down below. The violent movement, the vibration and the noise were bad enough on the bridge, but they must have felt worse for those cooped up within the vessel, unable to see or understand what was happening, so I had Eric, the chief steward, go around below to reassure and see to the welfare of those there.

Darkness fell on both the second and third nights as the onslaught continued, and the snow as thick as it can blow did so horizontally, but had no chance of building on the bridge windows, as it sometimes did, for the continual solid spray cleared them. The circular spinning clearview screens worked well, but by design there was not one on the centre line, as it would have interfered with vision for normal handling. However, I needed to remain there, because my hands hardly left the engine and course controls sited on the central console. Yet regardless of whether vision was obtained through window or clearview screen, only a wall of solid water, spray or snow, could be seen, and I think a mixture of sound, feel and instinct took over within me. Hardly a word was spoken, even as a half-cup of coffee and toast, which just amazingly materialised, were repeatedly placed in the housing on the fore part beneath the central window, which normally also held my binoculars and small cheroots. There was a professional tenseness in those on the bridge – perhaps some fear, I later learnt, as they stared ahead or into the radar. There were some icebergs around, but none big enough to shelter behind. However there would in that case have been some question as to whether to seek shelter behind one large enough to give us protection, for in these conditions the collapse of a large tabular berg into a multitude of smaller bergs and bergy bits would be likely, and this would have created too great a risk to us.

We mounted up and ran down each swell at angles I can only guess at, perhaps 45 degrees. I drank coffee, smoked an occasional small cigar, and grew very weary. I did not believe in sitting, and could not have done so safely, although my wife spent many hours silently observing from the pilot's chair, which was lashed in one corner of the bridge. She stayed silent, partly because she feared interrupting the concentration of those on the bridge, and also because her quiet voice was unlikely to be heard above the screaming of the storm and the thunderous crashing of sea against the hull. Her abiding memories of those days are of the monumental walls of sea, relentlessly roaring towards and towering above us; the blinding snow driving horizontally, and the mental vision of this cork-like speck being tossed upon a vast ocean, the latter part of this picture being brought about by movement that no human should have to endure, except at a fairground – and then only for minutes, not days.

I was not inclined to hand over to Stuart, my chief officer, as competent as he was. Several times he had taken command on previous voyages when I had gone home early, but this problem was my problem; it was I who was responsible for bringing our 100 souls safely through this. At the time I was almost instinctively not standing down, reflecting on the youngsters amongst the crew, the Fids who had only wanted to be transported safely home from the Antarctic, and the others down below, shocked, bewildered and frightened, even after the weather experienced during the season, by this attack upon their senses, this unbelievably violent world, this hell at sea. The enormity of the responsibility was stark. One false move during the whole episode could have brought about our demise. I say now that I earned all my wages of that season, during these and the next few hours. Never in my dozen years of command had the responsibility of it been highlighted so vividly.

I did not like to chance turning to run before the weather for fear of broaching. The vessel could go onto her beam ends and still right, but solid water entering through any hatch or doorway was of concern, although that should only be temporary, while she lay on her side. A more worrying scenario was that of water entering down the funnel and knocking out the electrical systems in the engine room, resulting in total failure of the engines. Also, once we were round and running before the weather, being pooped could not be discounted; this would also increase the possibility of water getting inboard and stalling the engine; lying a-hull without power was not something to be relished. The length between the swell tops was not that of a long rolling swell, but short for their height, so there would be little chance of turning 180 degrees, or anything approaching that, between one crest and another. Yet as we rode each swell, I watched and counted. Up, over and down, again and again. 'No,' I thought, 'I am not going to try it; the dangers are far too great.'

The wind then backed to the southward and I was told that the needle of the anemometer was once more almost constantly reading 120 knots, banging against the upper stud of the dial. There was soon a visible firming of floury ice around the wheelhouse windows, and as a glimmer of dawn filtered through on the fourth day, the same could be seen on most things forward, particularly the wires, rails and stanchions, growing measurably as we watched. Icing can get a grip in minutes. I asked the temperature, and it had dropped appreciably in the southerly wind. Many times before, until shelter was found behind a large berg or in an anchorage, I had seen the forecastle rails and stanchions grow with ice until they joined, turning that deck first into a swimming pool then into a multi-ton block of ice. Elsewhere on the ship a similar building of ice would begin. I had no knowledge of any ships capsizing from the build-up of ice in the Antarctic, shelter within pack ice or behind an iceberg or in the lee of islands usually being available, as opposed to the high incidence of this occurring in the Arctic, but here in mid-ocean there was nothing we could hide behind.

So, faced with ice forming rapidly overall, I had to turn. With my stomach in my mouth and hoping not to show any anxiety, I lit a cheroot and said, 'Warn everybody below that I am going to turn to run before it. She will roll like a pig until we are round, but then all will be well.' Then I waited for a while, but there was no discernible difference between swell lengths or heights; there was no obvious opportunity. As the ship neared a crest I finally applied full power and put her head fractionally off the wind. As the breaking sea smashed against the bow, I put the helm hard over, allowing the wind and sea atop the chosen swell to assist her rearing forefront to turn and be driven downwind, with the effect of the transverse thrust of the propeller helping. She sailed away like a bird. There was one gigantic roll, then a counter-roll, which I learnt tore the butcher's block from its welded deck brackets and smashed it through the galley bulkhead, but little major damage. None in the engine room, which I had feared; just a great many items, thought to have been well lashed, broken all about the ship.

We were round. The change from the violent struggle into the headwind and sea as they hurled their fury against us while we steadfastly maintained as near as possible our chosen hove-to course, was transformed into a poorly directed wallowing as the stern was lifted by each swell and as it ran under us we would surge forward. Each swell always seemed to come from one quarter or the other, skewing us off course, and looked almost certain to board us – but it never did, though our speed had to be continually adjusted to reduce the likelihood of being pooped.

I felt safe, and considered that I had won through. Then I cursed myself. Why had I not turned before? Over the next few hours the wind eased to a mere 80 knots, then veered westerly and down to 60; we were on an easterly course set to cope with the still enormous swell, running at a speed which appeared to drag the sea towards the stern but never actually aboard. The sky cleared, and I went out on to the bridge wing with my wife, Stuart joining us for a while. In the midst of the storms with the sky overcast with snow, we had felt, boxed up in the wheelhouse, that even during the hours of daylight our world had seemed so dark, overpowering, oppressive and dangerous. Now on the open wing, the sun shone and a carpet of spume sparkled as it drove past from stern to stem at about main deck level beneath us. Albatross, those graceful giants of the Southern Ocean, wheeled almost within arm's length, happily in close company with minute Wilson's petrels, both seemingly watching over the ship surging forward on each swell, as we stood pressed against the fore part of the bridge wing, hanging onto the dodger, exhilarated, and wondering how those creatures survived in such appalling conditions whilst we had barely done so. Our world had come alive again; our thoughts turned to matters beyond that of mere survival; the foreboding danger was gone and we were excitingly at peace. I went below and turned in.

The return

Some hours later Stuart called me to say that the weather had abated and he was more than happy to set our course for Stanley. Good sights had been obtained,[73] and as expected we were now several hundred miles to the east of our original course. A day and some hours later, after further fine weather and more sights to confirm our position, we raised Cape Pembroke lighthouse on the East Falklands. I frequently cut corners entering Port William, knowing it so well and often attempting to gain time to make a tide. But that morning I had already missed it, and caution seemed to be the order of the day after our recent thrashing. For my final entrance I also had a desire to linger, so took my time to soak up as much as I could of this place that had been our forward base and gateway to the Antarctic.

The Falklands were my second home, the place where, from my very first voyage as a youngster, so many Falkland Islanders and ex-pats had befriended me and welcomed me into their homes. They had afforded me shelter, tested my loyalties, placed me in dilemmas, tried my patience. Given me days of shooting hare over Mount Tumbledown, fishing for trout in Moody Brook, and riding over the camp. There were the short alliances with girls during my early visits, but who were invariably sensibly married to stable young local lads by the time I returned after the five-month elapse of the voyages to and from home and my leave. There were the parties aboard and ashore, and the drink-fuelled incidents like the marriage of the then shore-based Fid, when during the night we rolled a lawn mower up and down over the corrugated iron roof of the newlyweds' nuptial abode. And the baptism where both parents and vicar were so inebriated that the child to be named Bernard Brian, after slurred exchanges, was christened Bird-Brain. Ridiculous was the New Year's Ball at Government House when *Protector* officers inflated an enormous rubber boat on the dance floor, frightening the ladies out of their wits as they were squashed in their finery against the portrait-hung walls. Hilarious was a governor's wife who, during post-dinner party high jinks of indoor bowls, shouted the length of the room to the visiting head of the church in South America, 'Bishop, you've lost your balls!' Many memories. Captain Bill Johnston going down the gangway in his brown dressing gown and slippers for the last time; my steaming the *Bransfield* proudly up the harbour on her maiden voyage and slipping her alongside, just feet astern of my old *John Biscoe,* for their first meeting.

I could have taken a round turn out of her then, had there been room, to reminisce further. But I kept to the proper courses, a sensible distance off Cape Pembroke and Seal Rocks,

---

73 Astronomical observations.

rounding onto the westerly run that bisected them and Volunteer Point, which I had often visited to view the penguins; then up the harbour to the Narrows, the grasses on the Tussac Islands and the sweep of sand at Yorke Bay glistening after a passing shower, and Sparrows Cove to starboard, but without the SS *Great Britain*, she having been removed to Bristol. How many times in all weathers had I done it? Port 15, a gentle turn to pick up the leading marks through the Narrows, laying off a touch to the west for the wind. This time I had a helmsman rather than handling the ship myself from the console, as this allowed me to wander and to talk to those on the bridge, perhaps to suppress my emotion as I enjoyed that last inward passage.

Did I really wish to go ashore? Was this life at sea, tested 'beyond the boundaries', and the resulting satisfaction, not for me? Snapping out of it, it was time for another decision. It was half-tide as we ran up Stanley Harbour, and we would never be able to get alongside, but not a soul aboard, let alone me myself, wanted to remain at anchor, be bothered with boats, delayed, and drenched with the inevitable spray getting ashore. We had already pumped out the peak tanks and most other ballast in preparation for eventually taking the mud, so I thought, 'Let's give everybody ashore and aboard one last surprise.' I squared up to the rickety old Public Jetty and went in at a rate of knots, judging when to let go the anchor in preparation for hauling off, for another rain squall was obscuring my favourite transit marks abeam. I managed to drive the bow into the mud near the jetty's western corner, repaired after my clouting it years before, and canted her well to starboard so that she could be hauled alongside with the mooring lines as the tide rose. There was never any doubt in my mind that some contraption of a hanging gangway could not be rigged to span the gap, and happy faces were soon crossing it in each direction. Those officers not already there came to the bridge. I shook them all by the hand, kissed my wife, and went below. Feelings of success at the safe completion of another season down south, amidst thoughts of all those previous experiences, never to be repeated, surged within me, mixed with a great deal of apprehension as to the wisdom of my decision to 'go ashore'.

13 Epilogue

I was fortunate to have joined FIDS when there was still some contact with the heroic age of exploration. Sir Raymond Priestly, who was with both Scott and Shackleton, directed the Survey from 1955–58, in the absence of Fuchs. Sir James Wordie, Frank Debenham and Sir Alister Hardy were all alive, as well as other expedition members from that era, and their stories were fascinating and advice helpful. I was also lucky to have gone south in the transitional stage between those brave days of early exploration, with the mentality of making do with whatever was provided by limited finances, and a well-funded government-backed operation. Of being aboard the ships as they improved from being weak, underpowered, and lacking any sophisticated navigational or communication equipment, to ones of immense strength, with effective delivery of high power, and carrying radars for different usage, and echo sounders to plumb some of the oceans deepest trenches, and with the ability to benefit from instant worldwide contact. The BAS ships are now extremely sophisticated and I presume safer, and it would be disrespectful of me to belittle their performance and achievements.

Yet satellite navigation, in revolutionising this once fascinating art, has removed the enormous satisfaction that was to be gained from finding one's way through the waters of the world. Obtaining sights and laying off courses to compensate for wind, current and pack ice movement is now replaced by little more than programming computers and watching dials and displays. Do not those aboard the present ships miss the joy of making Halley Bay from South Georgia through 1,500 miles of pack ice without a single astro-fix? Are they not deprived of the satisfaction of having achieved a station relief by manhandling stores ashore in foul weather? Instead of struggling, slipping, and backachingly landing boxes on ice-strewn shores and up ice ramps, at Rothera containerloads of cargo are now landed on a quay, and at Halley giant vehicles deliver huge loads directly from within the ship, such that the vessels' stay there is only a fraction of the time required for my visits. Of course, it is all very laudably modern and efficient, but the downside must be that the romance of the pioneering days has gone. That which was once an expedition is no longer. I am told that now two problems for the crews aboard the BAS ships are boredom and obesity!

The ships

The RRS *Shackleton*, acquired by FIDS in 1955 and undoubtedly the poorest of my three Antarctic ships, surprisingly outlived both the *John Biscoe* and *Bransfield*. During her ten-year careful husbandry by David Turnbull from 1959 until she was paid off by FIDS in 1969 and replaced by the *Bransfield*, she underwent many dramas and near-disasters. For example, she suffered extensive damage from grounding when, shortly after leaving Detaille Islet Base, she returned there to collect a forgotten wheelbarrow required for some building work elsewhere, making it the most expensive wheelbarrow of all time! Underpowered, she was also pushed ashore twice when caught in heavy ice amongst the rocks bordering the French Passage, just north of the Argentine Islands. David attempted to use explosives to free her, but patience, the elements and his expertise eventually refloated her, none too badly damaged.

David's exploits during the eruption of Deception have already been described, and must rank as the highlight of her life, whereas the nadir must surely have been the damage sustained to put her into dry dock in Montevideo in mid-season when I was aboard under her ill-chosen captain. In 1969 she was transferred to NERC, until 1983. After I had looked at her with a view to her purchase by Trinity House to conduct surveys for the Corporation's commercial arm, but had found her not quite satisfactory, she was then sold for work in the North Sea, eventually being taken on by Guard Line in 1992. Renamed *Sea Profiler* for seismic survey work, she remained in their employ, though somewhat modified, until 2004. A remarkable 50-year life.

The RRS *John Biscoe*, first season 1955–56, continued to sail south for FIDS, then BAS, until 1991. Undergoing a major refit in 1983 to take on a more scientific and hydrographic role, she went to the Antarctic every season for 35 years and to my knowledge became the longest serving of all Antarctic vessels.

During the 1980s the *John Biscoe* had her nearest brush with disaster. Under the command of the very experienced Chris Elliott, making a voyage to Rothera through pack ice in a notoriously difficult area for ice, close to the Amiot Islets off Adelaide Island, she ran into difficulties for five days, then became freed. With conditions easing, the more powerful *Polar Duke*, on a long-term charter to the American National Science Foundation, having been to their Palmer Station, hove into sight, making a similar passage. Chris, against his own judgement, but with help at hand to expedite his own goal, accepted her offer and cut in astern of her, still expressing concern about proceeding during the next four days of severe weather, yet by then barely able to extract his ship. But both became beset, and were swept perilously close past one iceberg by driving pack ice. Chris then saw the opportunity to shelter in the vortex of water behind the berg, but needing help from the *Polar Duke* to get there, was denied it. With the shoals of the Amiot Islands close by, and in fear of his ship grounding and being rolled over by the pack ice, he laid out two anchors, abandoned ship and accepted a passage to safety with his crew aboard the beset *Polar Duke*. When conditions eased enough for her to get under way, a rendezvous was made with the powerful German icebreaker *Polar Stern*, and the crew transferred to her, she eventually returning them to *Biscoe*. Her anchors had held and she was lying where she had been abandoned, so was got under way and completed the season. She was finally sold to Greek Cypriots and she operated in the eastern Mediterranean until being scrapped in Turkey in 2004.

In 1982 RRS *Bransfield* was very nearly lost by her alternative master, running aground at full speed approaching Rothera, opening up her bottom plating and returning home on her tank tops.[74] Under the command of Stuart Lawrence, she endured several more dramatic falls

74 Steel ships have tanks built between the bottom hull plating and an inner deck. This is called the tank top, and forms the base of any lowest hold or space. With the hull ruptured it is possible to float on the tank tops.

of ice upon her whilst lying against the cliffs at Halley Research Station. I had begun to wonder, but later he even more so, whether it was not completely insane to place a ship in such a dire position. He also got into great difficulties in heavy pack ice in that same area off the Amiot Islands but he managed, in the classical manner, to hide in the vortex behind an iceberg, to avoid being crushed in ice or driven ashore. Her appalling end came about at the hands of Civil Service accountants, rather than at sea. After 29 years of service, and still in her prime, she was deemed unsuitable for the more scientific role then required of her and was sold to Rieber Shipping. They immediately scrapped her as part of the agreement for them to provide her chartered replacement, the *Polar Queen*, renamed RRS *Ernest Shackleton*, the intention presumably being to obviate her becoming competition if in the hands of another owner.

HMS *Protector* worked first with FIDS in 1955, was replaced by HMS *Endurance* in 1968 and then scrapped in 1970. The first HMS *Endurance*, formerly the Lauritzen Line's *Anita Dan*, did not have a problem-free life. Ably commanded by Peter Buchanan during her first two years, she later struck a rock surveying during the 1970s in Marguerite Bay where, on my and the *Shackleton's* first voyage we had run into such difficulties amongst shoals. In 1984 she ran aground near Brabant Island, and had to be towed off by *Bransfield*. Replaced after an interesting escape from an Argentinean submarine at South Georgia during the Falklands War, she was herself replaced by the second HMS *Endurance*, formerly Rieber's *Polar Circle*. A fine ship, she came to grief in the Magellan Straits, extensively flooding, when apparently inlet valves, being cleared, probably, of kelp, could not be shut. She had to be brought back to the UK aboard a transporter and at present, some years after the incident, awaits her fate in Portsmouth Harbour. In the meantime the *Polar Bjorn,* yet another Rieber vessel, has been chartered by the MOD and renamed HMS *Protector,* now providing the naval presence in the Antarctic.

Cruise ships can hardly be spoken of in the same context as the FIDS, BAS or navy vessels, but the fact is that unwisely, in my opinion, without better classification, control and supervision, many now visit Antarctica. This is now possible on account of the extensive hydrographic surveys having been undertaken over the years by BAS and naval ships, and the effects of global warming reducing the amount of pack ice in mid-season in the northern peninsula area. Yet many dangers still exist for the inexperienced, unwary and over-confident. It might appear that I resent their intrusion into my world, yet in truth they hardly scratch the surface of that domain, pushing only marginally into the 'Banana Belt', but the saga of their catastrophic performance underlines the threat of the larger vessels to the environment and the safety of the enormous numbers aboard.

The *Lindblad Explorer,* which I had attempted to salvage from King George Island in 1972, had already gone aground in the Gerlache Strait on her first voyage south in the previous season. Admittedly she made many short voyages to southern shores, on one rescuing the crew of the Argentine supply vessel, *Bahía Paraiso*, which caused a large environmental disaster when she grounded and capsized amongst the shoals off Arthur Harbour, Anvers Island, which had caused us so many difficulties in my early years. Renamed the *Explorer*, she sank in Bransfield Strait in November 2007, after hitting ice only 50 miles from where we had tried to pluck her off the rocks of King George Island. Miraculously good weather enabled all the passengers and crew to be rescued. The previous season and earlier that same season, two cruise vessels were blown ashore and damaged at Deception Island, one exactly where in 1964 we had witnessed the US Coastguard cutter, *East Wind*, driven onto the beach. In December of the same year the Norwegian *Fram* drifted onto a berg, and the next year, 2008, the *Ushuaia* was badly damaged on rocks in Wilhelmina Bay, off the Gerlache Strait. In 2009 the *Ocean Nova* was blown ashore near Stonington Island, in a Neny Howler.

BAS currently deploys two ships. The RRS *James Clark Ross* was purpose-built for them in 1991, to Lloyd's Ice Class 1A. The *JCR*, as she is known, is of 5,556 gross tons, 99 metres in length and 19 metres beam, and draws 6.3 metres. She has Wartsila Vasa diesels delivering 8,500 shaft horse power giving a maximum speed of 15 knots, with a 57-day range at a service speed of 12 knots, and has both bow and stern thrusters. She carries 36 crew and 41 supernumeraries, she is equipped with a variety of echo sounders for both navigational and scientific use, and her hull is configured to minimise ice and air becoming trapped beneath it, so as not to interfere with her performance. She has wet and dry laboratories, trawl and coring cables, an A-frame, a jib crane, and a gantry for seabed scientific work, sonar, sub-bottom profiler and acoustic Doppler current profiler. She is a fine multi-purpose scientific research and resupply vessel.

The RRS *Ernest Shackleton* is rather more of a workhorse. Having a crew of 22 and enabling 50 BAS personnel to be carried, she is of 80 metres in length, 17 metres beam and 6 metres draught, and has a range of 120 days. She has 3,000 cubic metres capacity for cargo, but a limited scientific fit. Of 7,200 shaft horsepower with a variable pitch propeller, she has three tunnel thrusters, and an Azimuth thruster, together providing her with exceptional manoeuvrability.

In mid-2014 it was agreed that a £200 million vessel 130 metres in length would be built soon to support the Survey. Whilst laudable, the intention eventually to operate this single ship is ill-conceived. With global warming, yet another Antarctic anomaly is that increased cold winds off the continent often occur, creating more pack ice, so now the ships are just as vulnerable as they were in my years, if not more so. As they are often many hundreds of miles within the pack, I believe assistance, or another vessel to take over their role whilst one is beset or disabled, to be vital. A much larger ship, as proposed, is not ideal for working pack ice, length being a criterion unless that ship is nuclear-powered. Nor, with its deep draught, would it be able to roam the shallower bays, anchorages and constricted station and refuge sites of the Antarctic Peninsula.

The Survey

BAS moved from London to a purpose-built establishment on a large site on the outskirts of Cambridge, its numbers growing from a handful in 1955 to the present-day figure of over 400. Although it was still an arm of the NERC, discussions took place as to its being merged with the National Oceanographic Centre in Southampton, also a component body of NERC. In the past few years the government has already stepped in to prevent NERC from closing one of the three remaining bases and disposing of one of the two ships.

Since the formation of FIDS in July 1945 under the dedicated, fiercely independent and politically adept Sir Vivian Fuchs as field commander at Stonington Island in 1947, and director from 1950, and his like-minded successor Dr Richard Laws, director from 1974 until 1987, it has been directed by a number of academics. Some, uncomfortable answering to their masters at NERC, have quickly moved on. NERC seems to overlook the international status of BAS, the importance of understanding the relevance of polar research to world science, and the requirement for it to be independent in order for it to perform its role within the Antarctic Treaty System, the only continent on the planet to be so regulated. Also overlooked is the influence which Britain exerts through BAS on all matters regarding the Southern Ocean and Antarctica, and the signal that the erosion of its identity would give to other nations, particularly the Argentine. Fortunately those with a better understanding of the matter forced the proposals of such a merger to be dropped, and the Survey has since early 2014 been under the directorship of an experienced and dedicated Antarctician and Polar medalist, Professor Jane Francis.

The Falkland Islands

Since my time in the Falklands, they have been transformed by the 1982 war, and the population doubled by the establishment of the garrison and airfield at Mount Pleasant. Also, sadly, many of the pleasurable areas I had known for shooting, fishing and walking are still out of bounds due to remaining mines. The BAS ships now use the rather forlorn Mare Harbour, developed to serve nearby Mount Pleasant, which cannot be the same for the crews as our visits to Stanley. The airport has facilitated incoming BAS personnel to join the ships in the Falklands and also enabled flights to and from the Antarctic to stage through it. The islands themselves have gone from a variable wool-sale-dependent economy to one first based on fishing, when in 1986 licence fees were introduced for squid fishing, then to the development of support services for such craft, then to tourist-based activities and now the beginnings of an oil industry. Personal wealth has rocketed and likewise through taxation of the various activities the islands are now self-supporting other than for defence.

Research stations (the bases)

Halley Bay, established in 1956 and rebuilt twice by 1973, has been rebuilt three more times since my last visit in 1974 and was renamed Halley Research Station in 1977. Some of the buildings of Halley VI are now built upon steel legs that are jacked up annually to raise them above the accumulating ice, and are also on skis so that they can be relocated to a position clear of the drift of the previous season and further from the continually calving ice front. The base operates with just under 20 wintering personnel, but in summer this can swell to around 70. The station's research, particularly the geophysical, is regarded as producing scientific data internationally recognised as of exceptional quality and value.

Rothera Research Station has developed to become the hub for the support of British Antarctic Field Science. Twenty-two wintering personnel swell to over 100 in the austral summer. Three de Havilland Twin Otter aircraft and one de Havilland Dash 7, operating from a 900-metre-long crushed stone runway, provide links with South America and the Falkland Islands, and support the field parties. A quay allows the supply vessels to berth alongside the station, which contains a large laboratory and supporting buildings housing accommodation, workshops, generators and a hangar. Geology, glaciology, meteorology and science of the upper atmosphere are carried out. Studies are also particularly made to observe the changes in species that in turn, indicate climate change. The scene at Rothera Point is now a far cry from those few footprints of our landing party from *Bransfield* upon that beautiful empty shore in 1974.

Signy Research Station, known as Signy Island Base until 1977, has operated only in the summer to conduct biology since the 1996–97 season. This base, together with Halley and Rothera now comprise the present BAS shore station activity in the Antarctic, with field parties ranging from the latter two. Research also continues at two sites on South Georgia.

The Argentine Islands base that opened in the spring of 1947 was occupied by FIDS then BAS for 47 years, the longest occupation of any British Antarctic Station to date. In 1996 it was transferred to the Ukraine. The Hope Bay base was transferred to Uruguay in 1997, and its satellite, View Point, was transferred to Chile in 1996, as was the original Adelaide Island base. The Brazilians cleared the site of the Admiralty Bay base and built their own station in its place. The Stonington Island and Horseshoe Island bases have both been closed, the former site cleared, the latter preserved as a depot.

My successors

Stuart Lawrence took command of the *Bransfield* when I retired in 1974, then the *Ernest Shackleton* in 1999. Chris Elliot became master of the *John Biscoe* in 1976 and subsequently of the *James Clark Ross,* on her launch in 1991. They both gave 30 years of adventurous yet safe, dedicated, diligent, highly successful and unblemished command. The outstanding service that these two master mariners gave to the British Antarctic Survey, and their incredible record and achievements, cannot be exaggerated, and consequently, neither can their contribution to the understanding of the Antarctic and its science which in turn contributes so greatly to our overall knowledge of the planet. In retirement, both acted as ice pilots for cruise ships, but unsurprisingly found the experience very frustrating; being able only to advise cruise ship masters, who were so different in their way of thinking, and their style of operating and command, neither of them did it for long.

Malcolm Phelps served as master of the *John Biscoe* for over 20 years, retiring when she was paid off. He too gave exemplary service in his quiet, dependable manner, bearing the brunt of some of the less glamorous aspects of the work south. Nick Beer, my second officer during the latter voyages, gained command of the *Bransfield* before joining the Marine Accident Investigation Bureau and attaining a senior position within it.

Myself

Before I formally resigned from BAS I sought advice as to the stability of Trinity House before agreeing to join it. An august body it may have been, serving the mariner almost without break since 1514, but at that time, in mid-1974, the Board was still concerned as to the question of liability regarding the marking of the wreck of the *Texaco Caribbean,* and the consequent collisions into it by the *Nikki* and the *Brandenberg* in the Dover Straits in 1971. I was also made aware by BAS that the strain on personnel of the nine-month Antarctic voyages was thought unreasonable, and that double manning[75] was to be introduced for senior officers. In my mind this was somewhat of a carrot to remain in their employ, but realistically I knew that I had seen the best years of Antarctic exploration with BAS, and knew also that although I should be required to sail south for only half a season, that I should have to put in time working in the office as a marine superintendent, which did not appeal. Further, it was proposed, quite sensibly, to utilise the ships when not on their Antarctic voyages by sending them to undertake oceanographic research in open northern waters, and to become involved in projects related to commercial oil; but neither of these ideas interested me. Perceiving the end of my role as I had known it, with a ship that was not my own but under shared command – and attracted by the prestige of Trinity House, the role it played in the nation's maritime affairs, and being at home with my wife, I left the Survey.

In July 1974 I became an Elder Brother of the Corporation of Trinity House, joining the Court and Board as its junior member. Commuting the daily 50-mile journey to and from home to the Tower Hill headquarters in London was somewhat of a culture shock for me, far removed from sprinting up the companionway in 30 seconds to the bridge when I wished, or was needed. Worse was entering the endless round of committee meetings with their laboriously reached joint decisions, so different from my unquestioned individual, satisfying and unique command. A short respite from my new life came about when the Board allowed me to fulfil an ambition to make a voyage under sail in the Southern Ocean. I had provided input to the committee of the 1977–78 Whitbread Round the World Race, become known to the crews when talking to them about the Southern Ocean, and was asked to join one on the Cape Town to Auckland leg of the race, during an extended summer leave. The Board believed it would improve their elderly image! It involved 42 days of good sailing, reaching 59 degrees latitude south of Australia, with a very small amount of loose ice thrown in for good measure. One of my principal memories of the passage, as of those in the *Bransfield*, was the outstanding ability of the '20'-year-olds aboard, in this case on the foredeck, changing sail, during heavy weather in the Southern Ocean.

Shortly after my return from that, I was given the task of leading the design team of the replacement for our flagship *Patricia*. I cannot say that I engineered it, or even influenced the decision, other than that I got along well with British Shipbuilders, but the order was placed with the Robb Caledon yard at Leith, where *Bransfield* had been built. For many months I was reunited with my friends in the yard. Many of the innovations of the *Bransfield*, particularly her bridge layout and control systems, were replicated in the new build. She is still deemed a most successful ship. Finance of shipping was a new subject area for me and exploring the funding for her partly led to further ventures, whilst the new THV *Patricia* was very cost-effectively leased, saving capital expenditure on the part of the General Lighthouse Fund.

I had come ashore not only to join Trinity House but also in the hope of furthering two ideas. One to promote Antarctic cruising in a manner I thought fit, with small, properly classified and manned ice-strengthened ships. The second, to build an ice-strengthened supply ship for charter to nations wishing to become involved in Antarctic research, thereby gaining a position within the Antarctic Treaty System. The cruising venture almost came to

---

75 Two individuals alternately occupying the same post.

fruition with the Russian-owned shipping line CTC, British Airways and myself as partners, the project collapsing only five days before its launch at the Earls Court Travel Show because Moscow raised fears at the last minute that their relations with the Argentine would suffer if the ships of the consortium were to sail via the Falklands.

The supply vessel venture came about on meeting a German shipowner with similar ideas to my own for a support vessel, when exploring the build and finance of *Patricia* on the Continent. We put together a partnership of professional personal investors attracting tax incentives, designed a 6,000-ton polar vessel, the MV *Icebird*, and built her at Oldenburg. She was the first vessel to be registered at Lloyd's with an asymmetric stern to improve her propulsive efficiency by some 10 per cent, to counterbalance the loss of a similar degree due to the ice-form cutaway bow. Built speculatively, but financed greatly by both federal and regional government loans through a consortium of banks, six months into her build she was chartered by the Australian Government, already signatories to the Antarctic Treaty, for their southern expeditions. She arrived at Casey Station on the Antarctic mainland barely 15 months after my first meeting that had led to her build.

Anticipating a replacement for HMS *Endurance*, and being aware of interest in polar ships from those nations wishing to become involved in the Antarctic, I formed a small shipbroking company in partnership with a colleague from a firm with whom I had been a consultant. We were successful in acting for Rieber Shipping with the sale of their *Polar Circle* to the MOD, to become the second naval ship with the name *Endurance*. A fine vessel, she was unfortunately totally disabled in the Magellan Straits and awaits her fate as a hulk.

My next involvement with ships and ice came about through joining the committee of the TransGlobe Expedition led by Sir Ranulph Fiennes, and initially managing their ship, the *Benjamin Bowring*, formerly the *Kista Dan*. When she was unable to meet up with and retrieve Sir Ran as he emerged from the North Pole on the tongue of ice that moves south from the Arctic pack in late summer between Spitzbergen and Iceland, I flew to Longyearbyen to assist. We took the ship to 86 north, only 20 miles from Ran, but he was disinclined to travel towards us, even though I was confronted by enormous floes, many miles across and too hard for our little ship to break. Running out of time – for I was once again utilising my summer leave from Trinity House – I too failed to retrieve him, but six weeks later the ice itself brought him south.

My final foray into ice was very minor but enjoyable; when in retirement I sailed with Sir Robin Knox-Johnston aboard the sail training vessel *Malcolm Miller*. We took 40 fare-paying passengers on a three-week cruise sponsored by the *Financial Times* north from Aberdeen to Spitzbergen. After a fine, breezy passage under full sail, on arrival off the islands we edged between the light floes, being careful not to puncture the hull or damage the striker beneath the bowsprit – a far cry from working ice in the Weddell Sea in *Bransfield*! I found the scenery of Spitzbergen greatly disappointing compared to the grandeur of the Antarctic, and likewise the paucity of wildlife compared with that of the shores of the Southern Ocean.

Now I am in retirement with my wife, who so fortunately was able to experience the magic that is Antarctica, enabling us to share such memories. We do so as we enjoy the calmer scene of the waters of a Devon estuary, and we take comfort in our view of a starboard hand mark correctly aligned for the course which lies ahead.